Reveal MATH®

Implementation Guide

Program Reviewers

Nora G. Ramirez, M.A.
Executive Secretary,
  TODOS: Mathematics for ALL
Mathematics Education Consultant
Tempe, AZ

Courtney Koestler, Ph.D.
Associate Professor and Director, Ohio Center
  for Equity in Mathematics and Science
The Patton College of Education
Ohio University
Athens, Ohio

Theodore Chao, Ph.D.
Associate Professor,
  Department of Teaching and Learning
College of Education and Human Ecology
The Ohio State University
Columbus, OH

Ferdinand Rivera, Ph.D.
Professor, Department of
  Mathematics and Statistics
College of Science
San Jose State University
San Jose, CA

Polina Sabinin, Ed.D
Associate Professor of Mathematics
Bartlett College of Science and Mathematics
Bridgewater State University
Bridgewater, MA

Sarah Smitherman Pratt, Ph.D.
Lecturer, LOC EdD Program
Baylor University
WACO, TX

mheducation.com/prek-12

Copyright © 2022 McGraw Hill

All rights reserved. No part of this publication may be reproduced or distributed in any form or by any means, or stored in a database or retrieval system, without the prior written consent of McGraw Hill, including, but not limited to, network storage or transmission, or broadcast for distance learning.

Send all inquiries to:
McGraw Hill
8787 Orion Place
Columbus, OH 43240

ISBN: 978-0-07-694684-6
MHID: 0-07-694684-3

Printed in the United States of America.

3 4 5 6 7 8 9 LWI 26 25 24 23

Contents

Reveal Math **Authors** 4

Program Guide 7
    Program Overview
    Program Components
    Unit Features
    Instructional Model
    Lesson Walk-Through
    Unit End Matter
    Student Agency
    Math is...
    Focus, Coherence, Rigor
    Effective Mathematical Teaching Practices
    Math Practices and Processes
    Social and Emotional Learning
    Language of Math
    Support for English Learners
    Number Routines
    Sense-Making Routines
    Math Language Routines
    Fluency
    Assessments
    Math Probe — Formative Assessment
    Differentiation Resources
    Targeted Intervention
    Professional Learning Resources
    Digital Experience
    Reporting
    Class Management Tools

Content Guide 68
    STEM Careers
    SEL Competencies Correlations
    Key Concepts and Learning Objectives
    Scope and Sequence

Reveal Math Authors

Annie Fetter
- Math Education Specialist at the 21st Century Partnership for STEM Education
- Founding Member, The Math Forum
- Workshop Leader and Developer for Key Curriculum Press, 1995–2013

Linda Gojak, M.Ed.
- Director (retired), Center for Mathematics and Science Education, Teaching, and Technology at John Carroll University (OH)
- Past President, National Council of Teachers of Mathematics (NCTM)
- Past President, National Council of Supervisors of Mathematics (NCSM)
- Past member, NCTM Board of Directors

Susie Katt, Ph.D.
- K–2 Mathematics Coordinator, Lincoln Public Schools, Lincoln, NE
- Special appointment lecturer, University of Nebraska–Lincoln
- Robert Noyce National Science Foundation Master Teaching Fellowship, University of Nebraska–Lincoln
- R. L. Fredstrom Leadership Award, Lincoln Public Schools

Nicki Newton, Ed.D.
- Educational consultant and speaker
- Former bilingual elementary and middle school teacher
- Graduate instructor, Columbia, CUNY, MCNY, Mercy College, Cambridge College
- Founder of Math Online PD Academy

John SanGiovanni, M.Ed.
- Coordinator of Elementary Mathematics, Howard County, Maryland
- Past President, Maryland Council of Supervisor of Mathematics
- Graduate Program Coordinator, Elementary Mathematics Instructional Leader program, McDaniel College (MD)
- NCTM Board of Directors
- Member, NCSM Board of Directors

Raj Shah, Ph.D.

- Founder, Math Plus Academy
- Founding member, The Global Math Project
- Affiliate, Math Teacher Circles, the Julia Robinson Math Festival
- R&D Engineering and Management, Intel Corporation, 1999–2008

Sharon Griffin, Ph.D.

- Professor Emerita of Education and Psychology at Clark University, Worcester, MA
- Author of *Number Worlds*: A PreK–8 prevention-intervention mathematics curriculum
- Principal Investigator on research grants to enhance mathematics learning and development for low-income students
- Member of the Education Directorate of the Organization of Economic Collaboration and Development and Advisory Board for Mind, Brain and Education Journal, Basil Blackwell
- Content advisor to WGBH and PBS for several children's TV series designed to promote math learning

Ralph Connelly, Ph.D.

- Professor Emeritus, Brock University
- NCTM Mathematics Education Trust Board
- NCSM Board of Directors
- Past President, Ontario Association for Mathematics Education (OAME)

Ruth Harbin Miles, Ed.S.

- Math Coach to rural, suburban, and inner-city school mathematics teachers
- Mary Baldwin University Adjunct Instructor, Staunton, Virginia
- K–12 Mathematics Coordinator, Olathe District Schools, Olathe, Kansas
- NCTM Board of Directors
- NCSM Board of Directors

Program Guide

Program Overview	8
Program Components	10
Unit Features	12
Instructional Model	18
Lesson Walk-Through	20
Unit End Matter	26
Student Agency	28
Math is	30
Focus, Coherence, Rigor	32
Effective Mathematical Teaching Practices	34
Math Practices and Processes	36
Social and Emotional Learning	38
Language of Math	40
Support for English Learners	42
Number Routines	44
Sense-Making Routines	46
Math Language Routines	48
Fluency	50
Assessments	52
Math Probe – Formative Assessment	54
Differentiation Resources	56
Targeted Intervention	58
Professional Learning Resources	60
Digital Experience	62
Reporting	66
Class Management Tools	67

Program Overview

Reveal the Mathematician in Every Student

Reveal Math®, a balanced elementary math program, develops the problem solvers of tomorrow by incorporating both inquiry-focused and teacher-guided instructional strategies within each lesson. *Reveal Math* uncovers the full potential in every student.

> *Reveal Math* **champions a positive classroom environment** centered on curiosity, connection, and social-emotional development.

We believe that student thinking and communication skills are just as important as content knowledge in creating the problem solvers of tomorrow. That is why *Reveal Math* spends the first unit of every grade level focusing on the classroom environment, from modeling mathematical ways of thinking to setting classroom expectations as a class. This serves as a base for reminders that have been integrated into each lesson.

We also believe that the more voice students have in their learning, the better the outcome. Integrated social and emotional development objectives and opportunities for daily reflection not only help build ownership of learning and behavior, but help ensure that all students feel safe in the classroom where they know their ideas about math are welcomed and respected.

> *Reveal Math* **offers a flexible lesson design** providing access to rigorous instruction with embedded teacher supports and scaffolds.

We believe that sense making and the exploration of mathematics through rich discourse and productive struggle is the foundation of high quality mathematics instruction. Every lesson launches with an activity designed to spark students' curiosity and to afford them time to make sense of problems without the pressure of getting the right answer.

The *Reveal Math* lesson builds from that sense-making and holds student exploration at the heart of the lesson. Each lesson offers teachers with two options to facilitate productive exploration: Activity-based or Guided Exploration, both of which are infused with effective teaching practices, language scaffolds, and rich prompts to guide understanding through discourse.

> ***Reveal Math* tailors classroom activities to student needs** through insightful assessment and purposeful, multi-modal differentiation.

We believe that differentiation should focus on the learning goal and not a student label. A student's need could change day to day. That is why differentiation is informed by a daily Exit Ticket. Based on students' scores, teachers can assign activities to reinforce students' understanding of the lesson's concept, continue to build proficiency, or extend and apply their thinking.

Reveal Math delivers insightful instruction with assessment and differentiation resources to tailor classroom activities to students' needs. Every assessment is designed with the purpose of informing instruction and connects to meaningful differentiation or intervention resources at point of use to make the most of instructional time. Item Analysis and Reporting help teachers quickly act on assessment data and pull appropriate resources to address student need. Teachers can choose from a range of differentiated resources to fit not just the learning needs of their students, but also preferences for grouping of students, and implementation options.

Thank you for choosing *Reveal Math*!
— The Reveal Math author team

Program Components

Teacher Resources

Implementation Guide

Teacher Edition, 2-volume

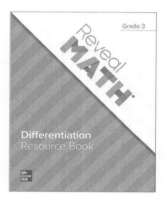
Differentiation Resource Book

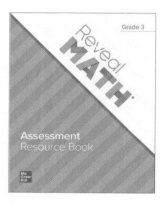
Assessment Resource Book

Teacher Digital Experience

Teachers have access to an intuitive and easy-to-use platform from which to plan and implement engaging instruction. The teacher experience includes:

- Daily interactive lesson presentations
- Engaging, rich differentiation resources
- Auto-scored practice and assessment items
- Customizable assessments and item banks
- Teacher and administrator data and reporting
- Professional development workshops and videos
- Unit and lesson files that can be downloaded with one click
- Ability to add resources, including presentations, website links, and more
- Classroom management and grouping tools

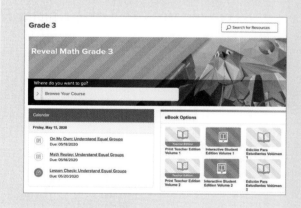

Workstation Kit

Application Station Cards

Workstation Teacher Guide

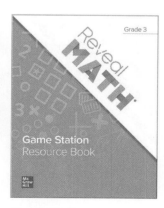
Game Station Resource Book

Student Resources

Student Edition, 2-volume

Student Practice Book

Spanish Resources

Student Edition, 2-volume

Student Practice Book

Differentiation Resource Book

Assessment Resource Book

Game Station Resource Book

Student Digital Experience

Students have access to a robust set of engaging digital tools and interactive learning aids, including:

- Interface designed for elementary students
- Interactive Student Edition
- Daily interactive practice with embedded learning aids
- Online assessments with interactive item types
- Digital games designed for purposeful practice
- Instructional mini-lessons to reinforce understanding
- Rich exploratory STEM Adventures
- Visual and dynamic WebSketch activities
- Animations, videos, and eTools

Application Station Cards

Program Components 11

Unit Features

Unit Planner
- Provides at-a-glance information to help teachers prepare for the unit
- Includes pacing; content, language, and SEL objectives; key vocabulary including math and academic terms; materials to gather; rigor focus; and standard(s).

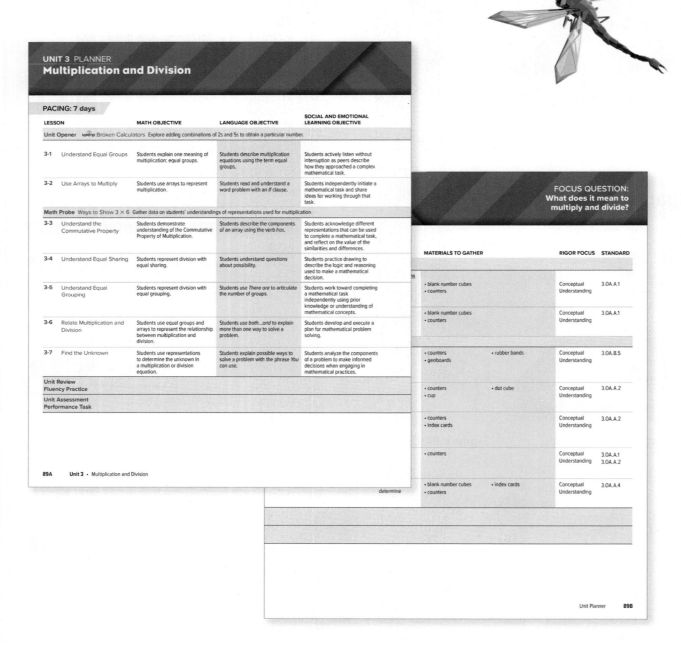

12 Implementation Guide

Unit Overview

The Unit Overview offers a comprehensive overview of the unit content for just-in-time professional support

Includes:

- content overview specifying focus, coherence, and rigor for the unit;
- pedagogical overview with NCTM's Effective Teaching Practices, Math Practices and Processes, and SEL competencies;
- language overview addressing language of math, math language development, and support for English learners; and
- unit routines: number routines, sense-making routines, and math language routines

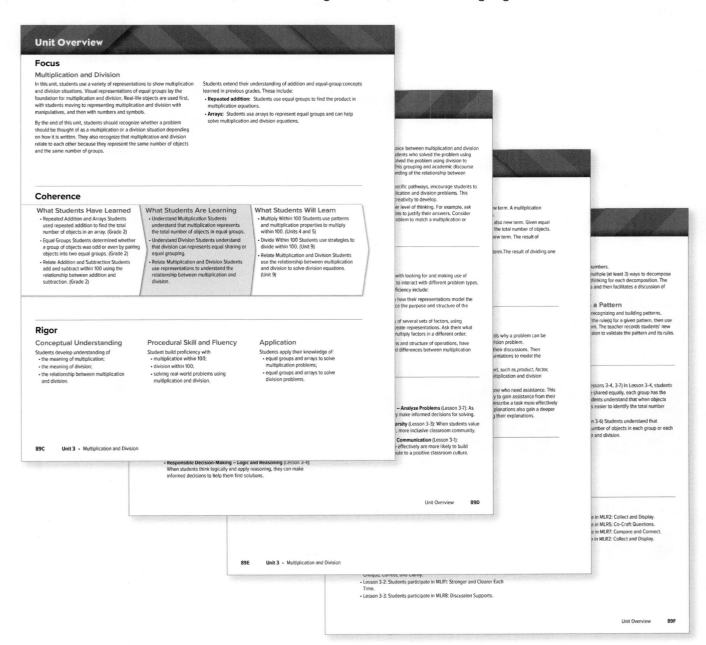

Unit Features 13

Unit Features

Readiness Diagnostic

- Offers teachers a unit diagnostic that can be administered in print or in digital. The digital assessment is auto-scored.
- Assesses prerequisite skills that students need to be successful with unit content.
- Item analysis lists DOK level, skill focus, and standard of each item.
- Item analysis also lists intervention lessons that teachers can assign to students or use in small group instruction.

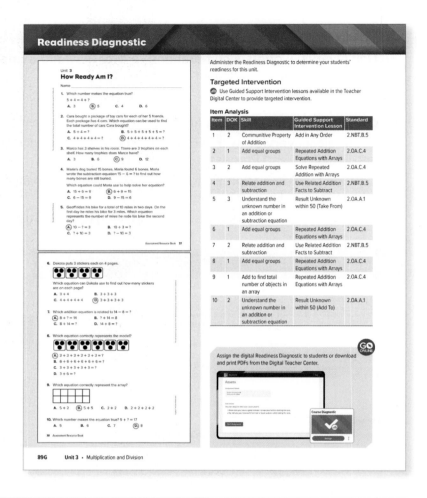

Guided Support Intervention lessons and Skills Support worksheets can be found with Unit resources in the Teacher Center.

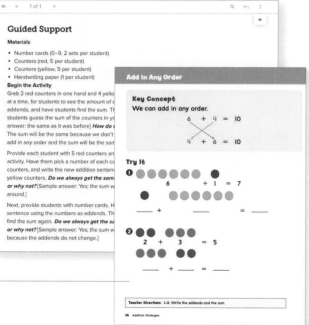

14 Implementation Guide

Unit Resources At-A-Glance and Additional Resources

- Offers an overview of differentiated activities from the Game Station, Digital Station, and Application Station that teachers can assign students;
- Also lists additional resources: Vocabulary Cards, Foldables, and Spiral Review

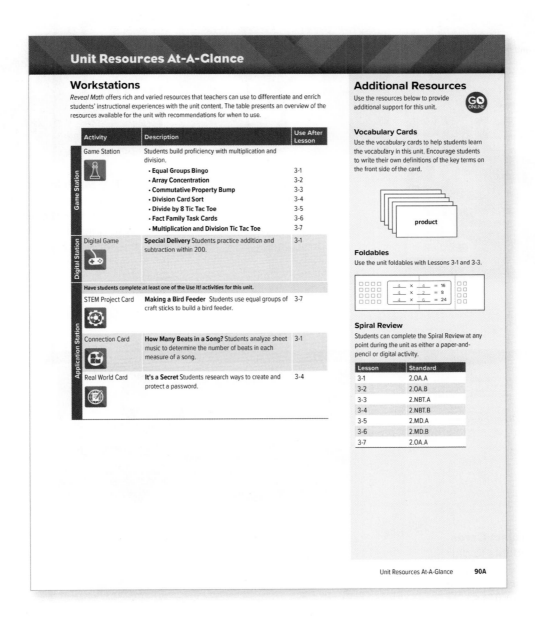

Unit Features

STEM-Focused Units

STEM-focused units highlight careers and real-world application of math to help students see the application and applicability of math to understand the world around them.

The **STEM Career Kid video** introduces a STEM career and provides an overview of the job responsibilities.

The **Math in Action** videos apply the unit math content to the STEM career focus to bring the content to the real world.

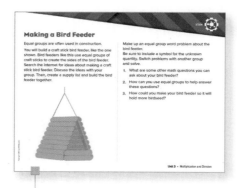

STEM Project Cards allow students to dig deeper creatively and apply their skills to learn more about the STEM focus within the unit.

STEM Adventures allow students to extend their thinking by applying skills to solve real-world problems through digital simulations.

16 Implementation Guide

Spark Student Curiosity Through Ignite! Activities

Each unit opens with an **Ignite!** activity, an interesting problem or puzzle that:

- sparks students' interest and curiosity,
- provides only enough information to open up students' thinking, and
- motivates them to persevere through challenges involved in problem-solving.

" Let's bring curiosity, wonder, and joy back into the classroom and make math irresistible for kids."

- Raj Shah, Author

Ignite! activities engage students in productive struggle as they provide only the information necessary to motivate and challenge the student.

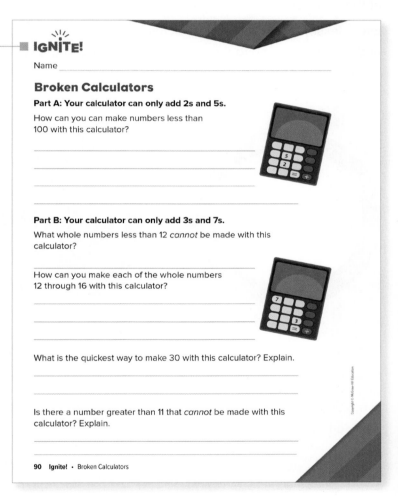

Learn more about Ignite! Activities with the Professional Learning Resources in your Teacher Center

Instructional Model

Reveal Math's lesson model keeps sense-making and exploration at the heart of learning.

Every lesson provides two instructional options to develop the math content and tailor the lesson to the needs and structure of the classroom.

Launch

Be Curious starts every lesson with the opportunity for students to be curious about math.

- Students focus on sense-making.
- Teachers foster students' ideas through meaningful discussion.

Explore and Develop

Explore and Develop unpacks the lesson content through either an activity-based exploration or guided exploration.

- Students explore the lesson concepts and engage in meaningful discourse.
- Teachers utilize effective teaching practices to help students make meaningful connections.

Practice and Reflect

On My Own offers students opportunities to engage with the math and reflect on their learning.

- Students practice lesson concepts by completing the On My Own exercise.
- Teachers have students reflect on the lesson content and their learning.

Routines

Instructional routines are embedded within every *Reveal Math* lesson to help students become proficient doers of mathematics.

Build Fluency

Number Routines
Support the development of flexibility with numbers and fluency with operations at the start of every lesson.

To learn more about Number Routines, see pages 44-45.

 MLR

Math Language Routines
Promote mathematical language use and development as part of math instruction.

To learn more about Math Language Routines, see pages 48-49.

Sense-Making Routines
Build sense-making as a foundation for problem-solving and mathematical modeling.

To learn more about Sense-Making Routines, see pages 46-47.

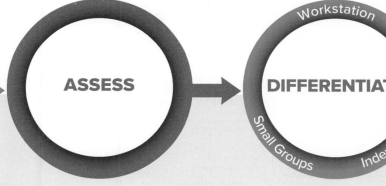

Assess

The **Exit Ticket** is a daily formative assessment to check for understanding.

- Students solve items related to the lesson content and reflect on their learning.
- Teachers use data to inform their daily differentiation.

Differentiate

Daily differentiation helps support every student in their path to understanding.

- Students work on differentiated tasks to reinforce their understanding, build their proficiency, and/or extend their thinking.
- Teachers work with small groups as needed.

Lesson Walk-Through

Lesson Overview

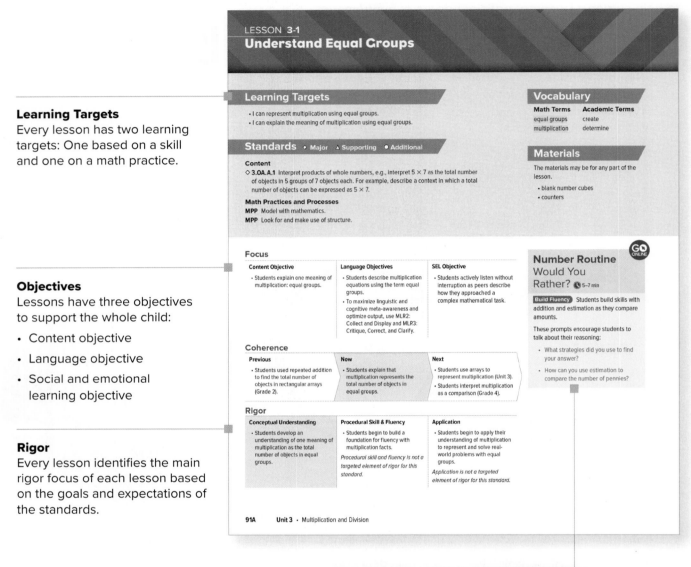

Learning Targets
Every lesson has two learning targets: One based on a skill and one on a math practice.

Objectives
Lessons have three objectives to support the whole child:
- Content objective
- Language objective
- Social and emotional learning objective

Rigor
Every lesson identifies the main rigor focus of each lesson based on the goals and expectations of the standards.

Daily Focus on Number Sense and Fluency

The Number Routine provides a daily focus on developing fluency and efficiency of strategy. The Number Routine can be completed at any point in the day to build number sense.

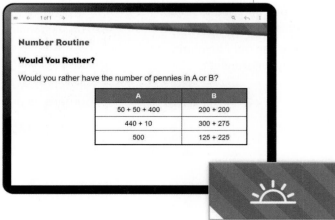

20 Implementation Guide

Launch

Sense-making routines launch every lesson, creating an equitable classroom culture where all ideas are welcome and respected. Student curiosity and ideas shared in Be Curious become the base for the day's instruction.

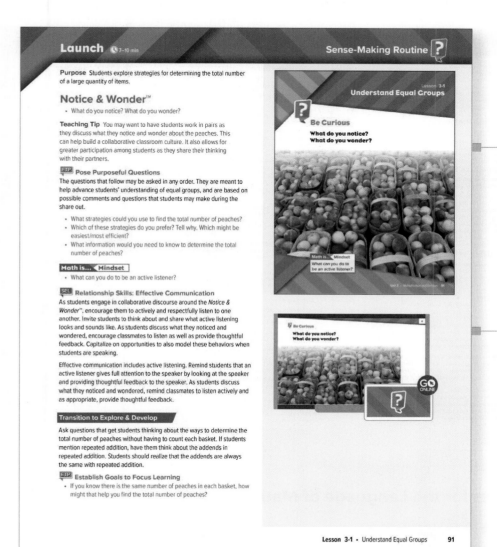

Be Curious offers a high-ceiling/low-floor activity with multiple entry points that allows every student to explore and discuss their ideas.

Math Is...Mindset prompts with teacher supports keep social and emotional learning at the top of students' minds as they interact and discuss throughout the lesson.

"All students have ideas about math that are valid and worth talking about."

— **Annie Fetter**
Contributing Author

Lesson Walk-Through

Explore & Develop

For the lesson's main instruction, the teacher can choose between two equivalent approaches to instruction, both of which provide the same level of access to rigorous content. Integrated Effective Teaching Practices guide instruction and discourse, keeping the student at the center of the learning.

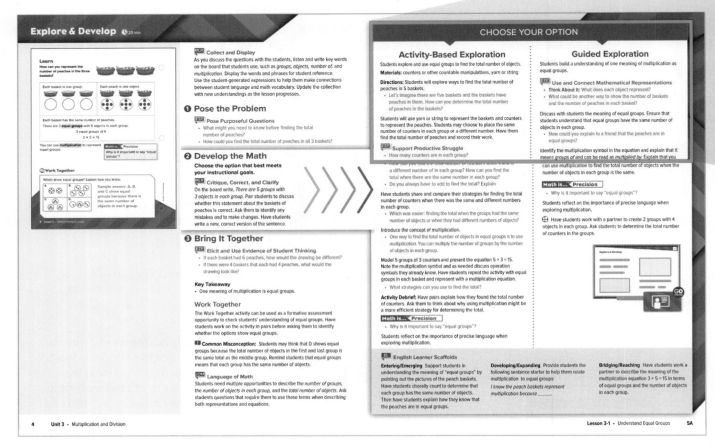

Comprehensive Supports for the Language of Math

EL — Built-in scaffolds that support English Learners as they interact with math language by speaking, listening, reading, and writing.

LOM — Language of Math supports how to talk about and think about math in context of the lesson content.

MLR — Math Language Routine promotes mathematical language use and development as part of math instruction.

22 Implementation Guide

CHOOSE YOUR OPTION

Activity-Based Exploration

Students explore and use equal groups to find the total number of objects.

Materials: counters or other countable manipulatives, yarn or string

Directions: Students will explore ways to find the total number of peaches in 5 baskets.

- Let's imagine there are five baskets and the baskets have peaches in them. How can you determine the total number of peaches in the baskets?

Students will use yarn or string to represent the baskets and counters to represent the peaches. Students may choose to place the same number of counters in each group or a different number. Have them find the total number of peaches and record their work.

ETP Support Productive Struggle
- How many counters are in each group?

Guided Exploration

Students build a understanding of one meaning of multiplication as equal groups.

ETP Use and Connect Mathematical Representations
- **Think About It:** What does each object represent?
- What could be another way to show the number of baskets and the number of peaches in each basket?

Discuss with students the meaning of equal groups. Ensure that students understand that equal groups have the same number of objects in each group.
- How could you explain to a friend that the peaches are in equal groups?

Identify the multiplication symbol in the equation and explain that it means *groups of* and can be read as *multiplied by*. Explain that you

Activity-Based Exploration allows students to explore concepts, develop and test hypotheses, and—most importantly—engage in productive struggle as they problem solve and generalize learning.

Guided Exploration offers a teacher-facilitated exploration with a question and answer format and collaboration to promote rich discourse about the lesson content.

Lesson Walk-Through 23

Lesson Walk-Through

Practice & Reflect

Practice and Reflect provides students with practice that address all elements of rigor.

On My Own exercises can be completed in the print Student Edition or the Interactive Student Edition.

Additional Practice Two pages of additional practice are available for every lesson. Students can complete these exercises either in the Student Practice Book or using the Interactive Additional Practice, which offers learning aids, such as hints, examples, and more.

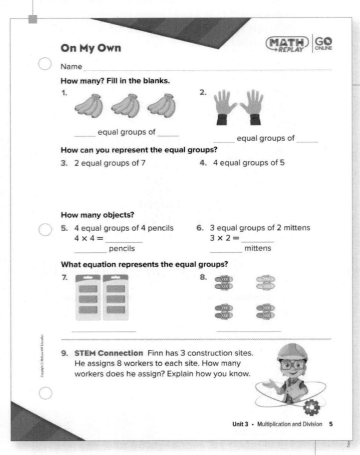

Math Replay videos
Students can access a lesson video to review lesson content as they complete practice assignments.

24 Implementation Guide

Assess & Differentiate

Every lesson closes with an Exit Ticket to check for student understanding and to serve as a source for recommendations for further differentiation.

Exit Ticket Teachers use students' scores on the Exit Ticket to decide on differentiated assignments from the robust differentiated resources available. When students complete the Exit Ticket in the digital environment, their work is auto-scored and mastery reports are generated.

Reflect On Your Learning allows students to reflect on their learning and communicate their confidence level with the teacher.

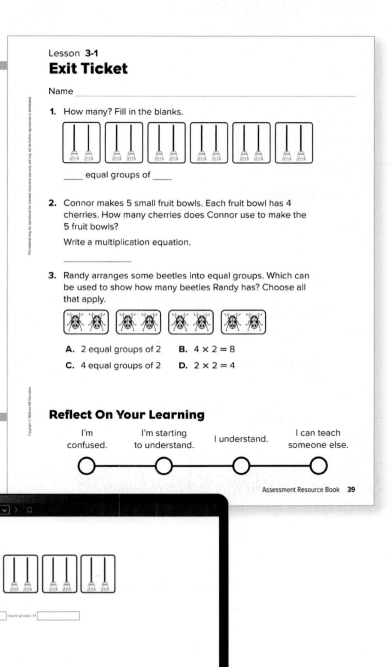

Lesson Walk-Through **25**

Unit End Matter

Unit Review

The Unit Review includes a vocabulary review, content review, and a practice performance task to get students ready for the unit assessment.

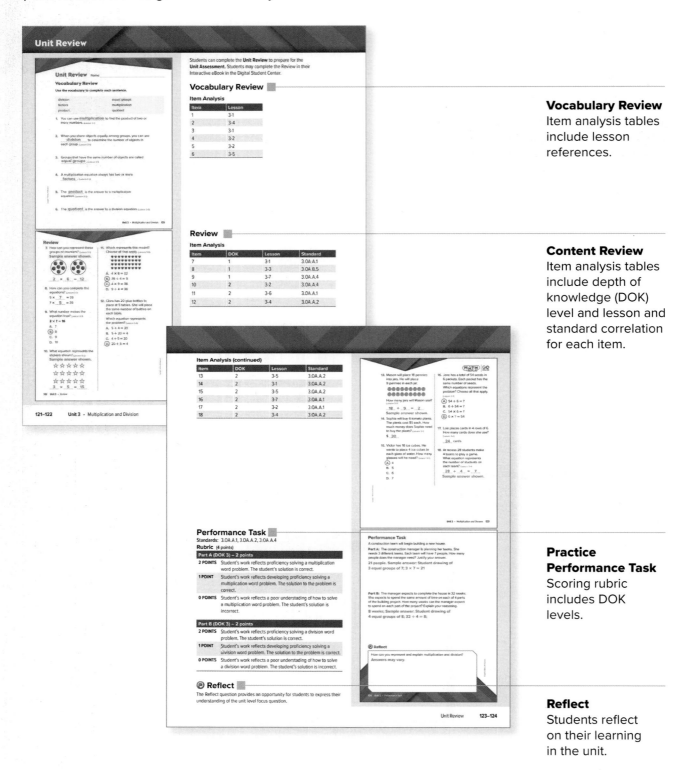

Vocabulary Review
Item analysis tables include lesson references.

Content Review
Item analysis tables include depth of knowledge (DOK) level and lesson and standard correlation for each item.

Practice Performance Task
Scoring rubric includes DOK levels.

Reflect
Students reflect on their learning in the unit.

Fluency Practice
- Includes fluency progression for each unit in the grade.
- Fluency expectations highlight expectations for current grade and previous grade.

Unit Assessments
- Includes scoring and rubric DOK levels for performance task
- Also includes item analysis tables with lesson and standard alignment and Guided Support Intervention lesson correlations to provide remediation.

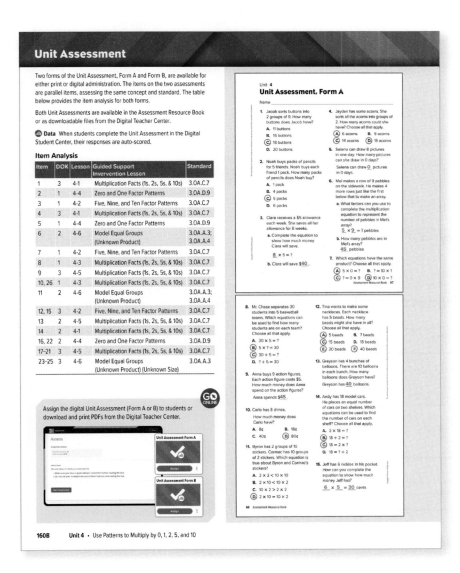

Student Agency

Giving students voice and choice is important for helping students feel ownership of and agency over their learning. *Reveal Math* offers learning activities designed to build student agency.

Math Is... Unit

The Math is... Unit is designed to build students' identity as doers of mathematics and to give them voice in their learning.

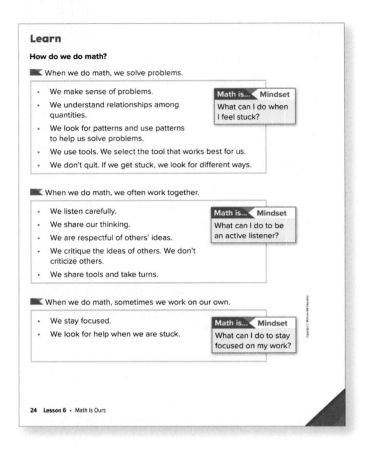

The Math Attitude Survey

Students build agency as they reflect on their self-perceptions and attitudes towards math. Students can review their responses periodically throughout the school year to track any changes in their attitudes or self-perceptions.

28 Implementation Guide

Growth Mindset

Students who believe that mistakes help them become better in math are willing to take learning risks. That strong identity and agency as doers of math leads to a growth mindset. The Math is... Unit encourages students to think of their math identify by:

- asking students to think about their math story, their math "superpowers," and their self-perception as "doers of mathematics."
- establishing what it means to be a "doer of mathematics."

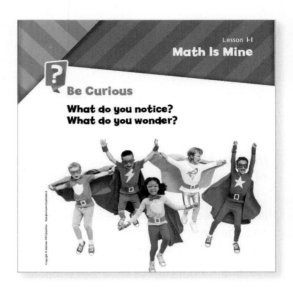

Productive Struggle

The **Activity-Based Exploration** offers problem-based activities that promote productive struggle as students explore concepts, test hypotheses, and formulate generalizations. Because students decide on strategies to use, they build student agency.

Directions: Students will explore ways to find the total number of peaches in 5 baskets.
- Let's imagine there are five baskets and the baskets have peaches in them. How can you determine the total number of peaches in the baskets?

Students will use yarn or string to represent the baskets and counters to represent the peaches. Students may choose to place the same number of counters in each group or a different number. Have them find the total number of peaches and record their work.

ETP Support Productive Struggle
- How many counters are in each group?
- How can you find the total number of counters when there is a different number of in each group? How can you find the total when there are the same number in each group?
- Do you always have to add to find the total? Explain

Metacognition: Daily Reflection

Reflection helps to drive accountability and the opportunity to think about thinking. Students reflect on both their understanding and behavior daily through **Reflect** prompts. Students also have the opportunity to reflect in the Unit Review and Math Probe.

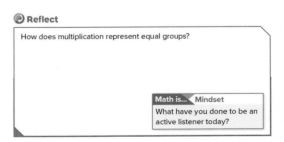

Student Agency 29

Math Is...

Ask your students, "What is math?" and you will likely get responses such as, "It's adding and subtracting." "It's multiplication and division." Some students might say, "It's solving problems," but their understanding of problem solving may be limited to doing textbook word problems. Taking the time to teach the Math Is... Unit is a critical step toward building a community of mathematicians.

Reveal Math **looks to change students' perceptions of themselves as doers of mathematics.**

The Math is... unit, the first unit in each grade, is designed to build students' mathematical identity and their agency as doers of math. The unit encourages students to:

- take ownership of their personal learning journey,
- see themselves as doers of mathematics,
- apply mathematical thinking to problem-solving,
- establish a productive and collaborative learning community and
- share ideas and collaborate freely.

Develop Student Identity
Lesson 1: Math Is... Mine

The first lesson aims to help all students see themselves as doers of mathematics and take ownership of their learning within the math classroom.

Students...
- learn about the teacher's personal math story,
- describe their math superpowers, and
- craft their personal math story.

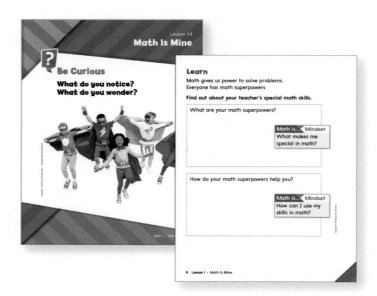

Learn more about the Math Is... Unit in the Professional Learning Quick Start in your Teacher Center.

Reveal MATH Quick Start
Select a topic to learn more.

30 Implementation Guide

Create Habits of Mathematical Thinking

Lessons 2–5: Math Is... a Way of Thinking

Lessons 2 through 5 focus on the mathematical habits of mind. Each lesson unpacks the thinking habits that are integral to problem solving and doing mathematics. Each lesson breaks down and models two common practices outlined by the mathematical practices.

Students...
- become mathematical thinkers,
- apply the math practices to problem solving, and
- communicate effectively about math.

Establish a Strong Classroom Community

Lessons 6: Math Is... Ours

In Lesson 6, students discuss what a positive and productive classroom environment looks like. Together, the class defines the classroom norms and expectations for the year. These norms help build a strong community.

Students...
- develop a voice and choice in their classroom environment.
- establish norms of interaction within the math classroom.

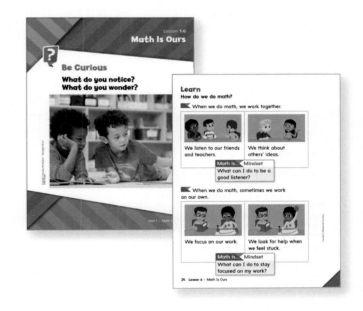

Math Is... Prompts

Math Is... prompts are embedded throughout the Student Edition. Math is... Mindset prompts found on the Be Curious page focus on building students' social and emotional competencies. Math is... prompts on the Learn page help students develop proficiency with the mathematical habits of mind. These prompts help students to own their learning journey throughout the year.

Math is... Mindset

How can you show others that you value their ideas?

Math Is... 31

Focus, Coherence, Rigor

Reveal Math was built to align to the three pillars of high quality mathematics curriculum — focus, coherence, and rigor.

Focus

The organization of units in *Reveal Math* puts the major content areas that are part of a pathway to algebra in the first half of the curriculum. Major content areas include:

- counting and cardinality,
- number sense with whole numbers, fractions, and decimals,
- place value with whole numbers and decimals,
- operations — addition, subtraction, multiplication, and division — with whole numbers, decimals, and fractions, and
- algebraic thinking.

Standards • Major ▲ Supporting ● Additional

Content
○ **2.0A.C.4** Use addition to find the total number of objects arranged in rectangular arrays with up to 5 rows and up to 5 columns; write an equation to express the total as a sum of equal addends.

Math Practices and Processes
MPP Construct viable arguments and critique the reasoning of others.

In the lesson overview in the Teacher Edition, the standards are labeled major, supporting, and additional.

This intentional organization of content gives students more time to develop deep understanding of and skill with these building-block concepts. Additionally, more opportunities are presented to apply these concepts in routine and non-routine situations.

Coherence

The scope and sequence of *Reveal Math* is built on the logical learning progression of mathematical content, connecting concepts across all grades and within each grade.

Coherence

What Students Have Learned	What Students Are Learning	What Students Will Learn
• Repeated Addition and Arrays Students used repeated addition to find the total number of objects in an array. (Grade 2)	• Understand Multiplication Students understand that multiplication represents the total number of objects in equal groups.	• Multiply Within 100 Students use patterns and multiplication properties to multiply within 100. (Units 4 and 5)
• Equal Groups Students determined whether a group of objects was odd or even by pairing objects into two equal groups. (Grade 2)	• Understand Division Students understand that division can represents equal sharing or equal grouping.	• Divide Within 100 Students use strategies to divide within 100. (Unit 9)
• Relate Addition and Subtraction Students add and subtract within 100 using the relationship between addition and subtraction. (Grade 2)	• Relate Multiplication and Division Students use representations to understand the relationship between multiplication and division.	• Relate Multiplication and Division Students use the relationship between multiplication and division to solve division equations. (Unit 9)

Unit-level coherence guidance helps teachers understand what prior knowledge students need to be able to access the unit content and for what math the current unit is building the foundation.

Coherence

Previous	Now	Next
• Students used repeated addition to find the total number of objects arranged in an array (Grade 2).	• Students understand that multiplication represents the total number of objects arranged in an array.	• Students learn the Commutative Property of Multiplication (Unit 3). • Students interpret multiplication as a comparison (Grade 4).

Lesson-level coherence guidance provides more granular analysis of the learning progression within a unit.

Rigor

Reveal Math was designed to focus with equal intensity on the three elements of rigor: conceptual understanding, procedural skill and fluency, and application. This focus allows students to develop deep and authentic proficiency of concepts.

Rigor

Conceptual Understanding	Procedural Skill & Fluency	Application
• Students develop understanding of one meaning of multiplication as the total number of objects in equal groups.	• Students begin to build a foundation for fluency with multiplication facts. *Procedural skill and fluency is not a targeted element of rigor for this standard.*	• Students begin to apply their understanding of multiplication to represent and solve real-world problems with equal groups. *Application is not a targeted element of rigor for this standard.*

Conceptual Understanding

The *Reveal Math* instructional model emphasizes sense-making as foundational to conceptual understanding.

- The **Be Curious** activity during the Launch focuses on sense making with different routines, notably, Notice and Wonder™.

During the **Explore & Develop**, instruction links the sense-making activity to conceptual understanding, making sure students understand the "why" behind operations and other important skills.

Procedural Skill and Fluency

Procedural fluency is built from a solid conceptual foundation. Lessons that focus on procedural fluency follow those that target conceptual understanding.

- **On My Own** exercises help students build procedural fluency as appropriate.
- **Fluency Practice** is designed to build students' fluency with operations.

Application

Students encounter real-world problems throughout each lesson. The **On My Own** exercises include rich, application-based question types, such as "Find the Error" and "Extend Thinking."

Daily differentiation provides opportunities for application through the Application Station Cards, STEM Adventures, and WebSketch Explorations.

The unit performance task found in the Student Edition offers another opportunity for students to solve non-routine application problems.

Effective Mathematical Teaching Practices

Reveal Math's instructional design integrates the **Effective Mathematics Teaching Practices** from the National Council of Teachers of Mathematics (NCTM). These research-based teaching practices were first presented and described in NCTM's 2014 work *Principles to Action: Ensuring Mathematical Success for All.*

These eight practices are:
- Establish mathematical goals to focus learning.
- Implement tasks that promote reasoning and problem-solving.
- Use and connect mathematical representations.
- Facilitate meaningful mathematical discourse.
- Pose purposeful questions.
- Build procedural fluency from conceptual understanding.
- Support productive struggle in learning mathematics.
- Elicit and use evidence of student thinking.

In each unit overview, teachers are presented with an unpacking of one of the teaching practices with suggestions for the teacher on successful implementation of the highlighted practice into instruction.

ETP Effective Teaching Practices

Implement Tasks That Promote Problem Solving and Reasoning

Students need to be fully engaged in a complex problem or task and be able to discuss it with someone before they feel they have fully grasped the concept. This is especially true in mathematics because there are often multiple ways to arrive at the same solution. Discussions with others allow students to discover varied points of view and different strategies that they can apply to future problems.

Problems that best promote reasoning and problem solving are non-routine problems, or problems that require a higher level of thinking. Multiple steps may be involved in solving the problem, which would allow for even more variety of strategies to be developed.

Students may have differing opinions or may be confused by the information provided during some of these lessons. When this occurs, spend time discussing these problems.

- When students are given the choice between multiplication and division in this unit, intentionally pair students who solved the problem using multiplication with those who solved the problem using division to analyze each other's answers. This grouping and academic discourse will allow for a deeper understanding of the relationship between multiplication and division.
- Instead of specifying tools or specific pathways, encourage students to find multiple solutions to multiplication and division problems. This allows for more strategies and creativity to develop.
- Assign tasks that require a higher level of thinking. For example, ask students to create representations to justify their answers. Consider having students write a word problem to match a multiplication or division equation.

Throughout the lessons are elements that embody each of the eight teaching practices. Look for the ETP icon.

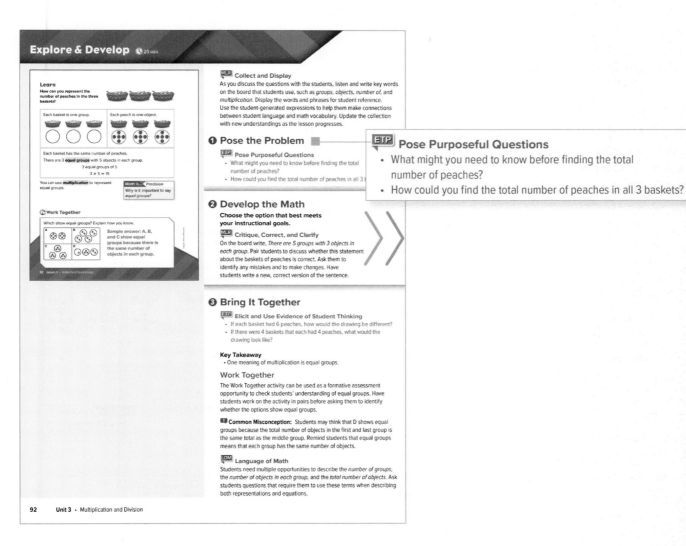

Effective Mathematical Teaching Practices **35**

Math Practices and Processes

To think like mathematicians, students must build habits of mind that help them develop a problem-solving frame of mind.

Reveal Math helps students build proficiency with these important thinking habits and problem-solving skills through the **Math is...** prompts found in every Learn. These prompts model the kinds of questions students can ask themselves to become proficient problem solvers and doers of math.

In the **Math is... Unit**, students are first introduced (or re-introduced) to the Math is... prompts.

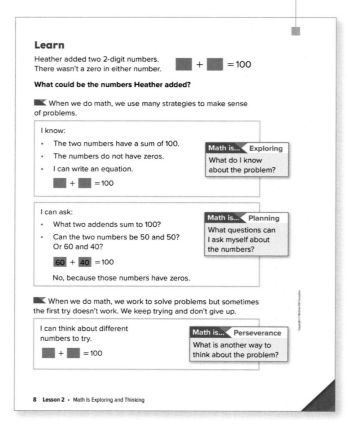

The **Math is...Prompt** in each Learn focuses on a different mathematical habit of mind.

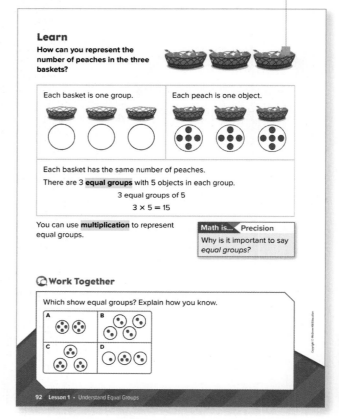

The Math is...prompts are also included in the Teacher Edition in both the Activity-based Exploration and the Guided Exploration.

CHOOSE YOUR OPTION

Activity-Based Exploration

Students explore and use equal groups to find the total number of objects.

Materials: counters or other countable manipulatives, yarn or string

Directions: Students will explore ways to find the total number of peaches in 5 baskets.
- Let's imagine there are five baskets and the baskets have peaches in them. How can you determine the total number of peaches in the baskets?

Students will use yarn or string to represent the baskets and counters to represent the peaches. Students may choose to place the same number of counters in each group or a different number. Have them find the total number of peaches and record their work.

ETP Support Productive Struggle
- How many counters are in each group?
- How can you find the total number of counters when there is a different number of in each group? How can you find the total when there are the same number in each group?
- Do you always have to add to find the total? Explain

Have students share and compare their strategies for finding the total number of counters when there was the same and different numbers in each group.
- Which was easier: finding the total when the groups had the same number of objects or when they had different numbers of objects?

Introduce the concept of multiplication.
- One way to find the total number of objects in equal groups is to use *multiplication*. You can multiply the number of groups by the number of objects in each group.

Model 5 groups of 3 counters and present the equation 5 × 3 = 15. Note the multiplication symbol and as needed discuss operation symbols they already know. Have students repeat the activity with equal groups in each basket and represent with a multiplication equation.
- What strategies can you use to find the total?

Activity Debrief: Have pairs explain how they found the total number of counters. Ask them to think about why using multiplication might be a more efficient strategy for determining the total.

Math is... Precision
- Why is it important to say "equal groups"?

Students reflect on the importance of precise language when exploring multiplication.

EL English Learner Scaffolds

Entering/Emerging Support students in understanding the meaning of "equal groups" by pointing out the pictures of the peach baskets. Have students chorally count to determine that each group has the same number of objects. Then have students explain how they know that the peaches are in equal groups.

Developing/Expanding Provide students the following sentence starter to help them relate multiplication to equal groups:
I know the peach baskets represent multiplication because _____.

Bridging/Reaching Have students work a partner to describe the meaning of the multiplication equation 3 × 5 = 15 in terms of equal groups and the number of objects in each group.

Guided Exploration

Students build a understanding of one meaning of multiplication as equal groups.

ETP Use and Connect Mathematical Representations
- **Think About It:** What does each object represent?
- What could be another way to show the number of baskets and the number of peaches in each basket?

Discuss with students the meaning of equal groups. Ensure that students understand that equal groups have the same number of objects in each group.
- How could you explain to a friend that the peaches are in equal groups?

Identify the multiplication symbol in the equation and explain that it means *groups of* and can be read as *multiplied by*. Explain that you can use multiplication to find the total number of objects when the number of objects in each group is the same.

Math is... Precision
- Why is it important to say "equal groups"?

Students reflect on the importance of precise language when exploring multiplication.

Have students work with a partner to create 2 groups with 4 objects in each group. Ask students to determine the total number of counters in the groups.

Math is... Precision
- Why is it important to say "equal groups"?

Students reflect on the importance of precise language when exploring multiplication.

Have students work with a partner to create 2 groups with 4 objects in each group. Ask students to determine the total number of counters in the groups.

Math is... Precision
- Why is it important to say "equal groups"?

Students reflect on the importance of precise language when exploring multiplication.

Math Practices and Processes

Social and Emotional Learning

In addition to academic skills, schools are also a primary place for students to build social skills. When students learn to manage their emotions and behaviors and to interact productively with classmates, they are more likely to achieve academic success. Research has shown that a focus on helping students develop social and emotional skills improves not just academic achievement, but students' attitudes towards school and prosocial behaviors (Durlak et al., 2011). A focus on social and emotional learning helps drive a positive math classroom where students are encouraged and motivated to engage in mathematics.

Reveal Math uses the SEL framework from the Collaborative for Academic, Social, and Emotional Learning (CASEL) to structure its SEL support. The five competencies —self-awareness, self management, social awareness, relationship skills, and responsible decision-making—are integrated throughout each grade level.

Each lesson has an SEL objective that is found in both the Unit Planner and the Lesson Overview page.

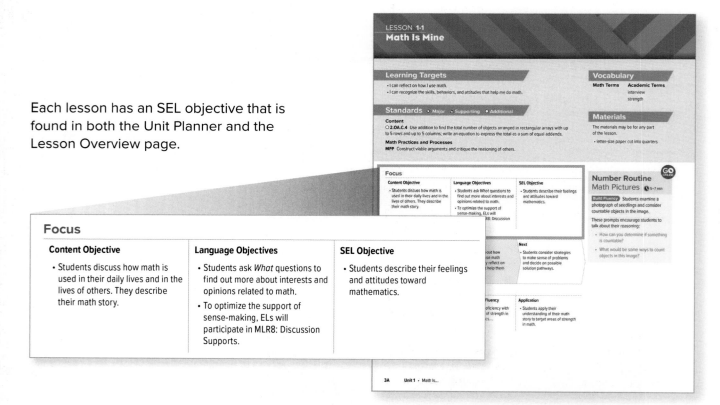

Durlak J, Weissberg, R, Dymnicki, A, Taylor, R, and Schellinger, K. (2011) *The Impact of Enhancing Students' Social and Emotional Learning: A Meta-Analysis of School-Based Universal Interventions.* Child Development. 82 (1).

Math Is... Mindset

The Math is...Mindset prompts have an intentional focus on the SEL competencies. The prompts are designed to help students build proficiency with the competencies through discussion and reflection.

Students first encounter the Math Is...Mindset prompts in Lessons 1 and 6 of Unit 1.

- Lesson 1 prompts help build students' self-awareness and self-management as they think about their attitudes towards and their strengths in math.
- Lesson 6 prompts focus on social awareness and relationship skills as students think about and discuss classroom norms for a productive learning environment.

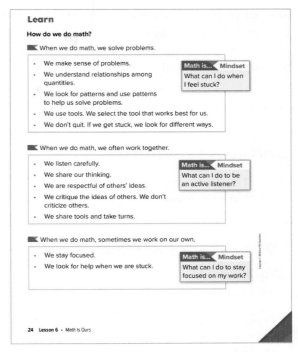

Starting in Unit 2, students encounter Math Is...Mindset prompts at the beginning and end of each lesson. These prompts focus on the same SEL competency. The prompt at the end of the lesson encourages students to reflect on their behaviors and actions related to the SEL competency during the lesson.

Math is... Mindset
- What can you do to be an active listener?

SEL Relationship Skills: Effective Communication
As students engage in collaborative discourse around the *Notice & Wonder*™, encourage them to actively and respectfully listen to one another. Invite students to think about and share what active listening looks and sounds like. As students discuss what they noticed and wondered, encourage classmates to listen as well as provide thoughtful feedback. Capitalize on opportunities to also model these behaviors when students are speaking.

Effective communication includes active listening. Remind [students that an] active listener gives full attention to the speaker by looking [at the speaker] and providing thoughtful feedback to the speaker. As stud[ents share] what they noticed and wondered, remind classmates to li[sten and,] as appropriate, provide thoughtful feedback.

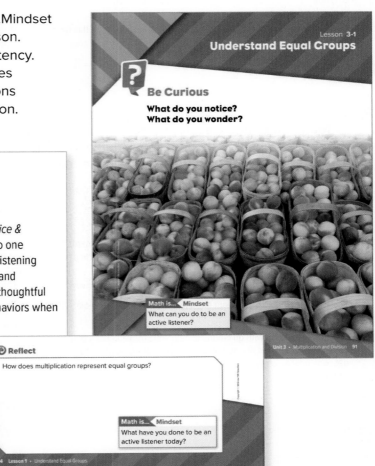

Social and Emotional Learning **39**

Language of Math

Reveal Math was developed around the belief that mathematics is not just a series of operations, but a way of communicating and a way of thinking. Some even argue that math is its own language, so to help students be successful in math, they need to learn the language of math.

Throughout *Reveal Math*, teachers will find language supports embedded to help students build a shared language and communicate effectively about math.

Unit-level Features

The **Language of Math** feature highlights math terms that students will use during the unit. New terms are highlighted in yellow. Terms that have a math meaning different from everyday means are also explained.

🗨 Language of Math

Vocabulary

Students will be using these key terms in this unit:

- **Array** (Lessons 3-2, 3-3, 3-6) Students encountered this term in the context of counting in Grade 2. Arrays with equal rows and columns are also used to represent multiplication problems.
- **Division*** (Lessons 3-4, 3-6) This is a new term. Division helps distribute objects equally among groups.
- **Equal groups** (Lessons 3-1, 3-6) Students were introduced to this term in the context of counting in Grade 2. Now they will apply that prior knowledge to multiplication.
- **Factor*** (Lessons 3-2, 3-3) This is also new term. A multiplication equation always has two (or more) factors.
- **Multiplication*** (Lessons 3-1, 3-6) This is also new term. Given equal groups, multiplication can be used to find the total number of objects.
- **Product*** (Lessons 3-2, 3-3) This is also new term. The result of multiplying one number by another.
- **Quotient*** (Lesson 3-5) This is also new term. The result of dividing one number by another.

Math Language Development This feature targets one of the four language skills—reading, writing, listening, speaking—and offers suggestions for helping students build proficiency with these skills in the math classroom.

🗨 Math Language Development

A Focus on Speaking

When speaking about mathematics, there are often complex concepts and processes to describe. There may be multiple steps or strategies in a problem and a variety of ways to explain similar processes, so mathematical explanations can be challenging for students to convey.

You can help students speak about math to partners or to the whole class by:

- Prompting students with questions to help start conversations or to gain a deeper level of discussion, such as prompting students to describe how an array can represent multiplication in real-world situations.
- Having students restate in their own words why a problem can be written as both a multiplication and a division problem.
- Providing students with visuals to aid in their discussions. Then students can describe how to use representations to model the problem.
- Having students use vocabulary in context, such as *product, factor, multiply,* and *divide,* while discussing multiplication and division situations.
- Pairing more advanced students with those who need assistance. This gives struggling students the opportunity to gain assistance from their peers. Sometimes students are able to describe a task more effectively to a classmate. Students who provide explanations also gain a deeper understanding of concepts while forming their explanations.

Lesson-level Features

The **Language of Math** feature promotes the development of key vocabulary terms that support how we talk about and think about math in the context of the lesson content.

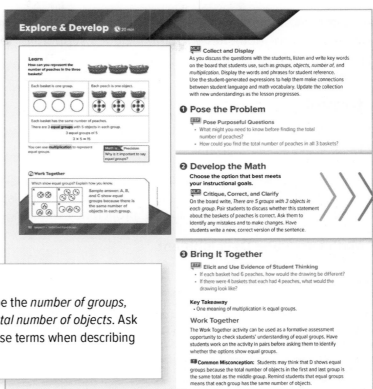

> **LOM Language of Math**
>
> Students need multiple opportunities to describe the *number of groups,* the *number of objects in each group,* and the *total number of objects.* Ask students questions that require them to use these terms when describing both representations and equations.

Support for English Learners

Ensuring that all students can achieve academic success is one of the guiding principles of *Reveal Math*. *Reveal Math* includes support for teachers to help English Learners achieve academic success.

Unit-level support

At the unit level are three features that provide support for teachers as they prepare to teach English Learners.

The **Math Language Development** feature offers insights into one of the four areas of language competence — reading, writing, listening, and speaking — and strategies to build students' proficiency with language.

MLD Math Language Development
A Focus on Speaking

When speaking about mathematics, there are often complex concepts and processes to describe. There may be multiple steps or strategies in a problem and a variety of ways to explain similar processes, so mathematical explanations can be challenging for students to convey.

You can help students speak about math to partners or to the whole class by:

- Prompting students with questions to help start conversations or to gain a deeper level of discussion, such as prompting students to describe how an array can represent multiplication in real-world situations.
- Having students restate in their own words why a problem can be written as both a multiplication and a division problem.
- Providing students with visuals to aid in their discussions. Then students can describe how to use representations to model the problem.
- Having students use vocabulary in context, such as *product, factor, multiply,* and *divide,* while discussing multiplication and division situations.
- Pairing more advanced students with those who need assistance. This gives struggling students the opportunity to gain assistance from their peers. Sometimes students are able to describe a task more effectively to a classmate. Students who provide explanations also gain a deeper understanding of concepts while forming their explanations.

The **English Language Learner** feature provides an overview of the lesson-level support.

EL English Language Learner

In this unit, students are provided with a number of scaffolds to support their comprehension of the language used to present and explain multiplication and division. Because many of the words and phrases used are likely unfamiliar to ELs, students are supported in understanding and using these words.

Lesson 3-1 – *equal*
Lesson 3-2 – *enough*
Lesson 3-3 – *any order*
Lesson 3-4 – *share, sharing*
Lesson 3-5 – *like*
Lesson 3-6 – *repeated*
Lesson 3-7 – *representation*

The **Math Language Routines** feature consists of a listing of the Math Language Routines found in each lesson of the unit.

MLR Math Language Routines

Mathematical Language Routines used in this unit give teachers a structured, yet adaptable format for amplifying and developing students' social and academic language. These routines can also be used as formative assessment opportunities as students develop proficiency in English and mathematical language. They can be used in ways that support real-time-, peer-, and self-assessment. For more information on the Math Language Routines, see the Appendix.

- Lesson 3-1: Students participate in MLR2: Collect and Display and MLR3: Critique, Correct, and Clarify.
- Lesson 3-2: Students participate in MLR1: Stronger and Clearer Each Time.
- Lesson 3-3: Students participate in MLR8: Discussion Supports.
- Lesson 3-4: Students participate in MLR2: Collect and Display.
- Lesson 3-5: Students participate in MLR5: Co-Craft Questions.
- Lesson 3-6: Students participate in MLR7: Compare and Connect.
- Lesson 3-7: Students participate in MLR2: Collect and Display.

42 Implementation Guide

Lesson-level support

Language Objectives

In addition to a content objective, each lesson has a language objective that identifies a linguistic focus of the lesson for English Learners. The language objective also identifies the Math Language Routines of the lesson.

Focus

Content Objective	Language Objective	SEL Objective
• Students explain one meaning of multiplication: equal groups.	• Students describe multiplication equations using the term equal groups. • To maximize linguistic and cognitive meta-awareness and optimize output, use MLR2: Collect and Display and MLR3: Critique, Correct, and Clarify.	• Students listen actively to classmates sharing their thinking. • Students actively listen without interruption as peers describe how they approached a complex mathematical task.

English Learner Scaffolds

English Learner Scaffolds provide teachers with scaffolded instruction to help students make meaning of math vocabulary, ideas, and concepts in context. The three levels of scaffolding within each lesson — Entering/Emerging, Developing/Expanding, and Bridging/Reaching are based on the 5 proficiency levels of the WIDA English Language Development Standards. With these three levels, teachers can scaffold instruction to the appropriate level of language proficiency of their students.

 English Learner Scaffolds

Entering/Emerging Support students in understanding the meaning of "equal groups" by pointing out the pictures of the peach baskets. Have students chorally count to determine that each group has the same number of objects. Then have students explain how they know that the peaches are in equal groups.

Developing/Expanding Provide students the following sentence starter to help them relate multiplication to equal groups:
I know the peach baskets represent multiplication because ____.

Bridging/Reaching Have students work a partner to describe the meaning of the multiplication equation 3 × 5 = 15 in terms of equal groups and the number of objects in each group.

Math Language Routines

Each lesson has at least one Math Language Routine specifically designed to engage English Learners in math and language. See pages 48-49 for more information.

 Collect and Display
As you discuss the questions with the students, listen and write key words on the board that students use, such as *groups, objects, number of,* and *multiplication*. Display the words and phrases for student reference. Use the student-generated expressions to help them make connections between student language and math vocabulary. Update the collection with new understandings as the lesson progresses.

Support for English Learners

Number Routines

Routines play an important part in the *Reveal Math* instructional model. Routines help students establish expectations in terms of behaviors or thinking processes. Routines also communicate to students what is important to know and do. The frequency of routines helps to build fluency with the habits that the routines look to instill in students.

The **Number Routines** in *Reveal Math*, authored by John SanGiovanni, are designed to build students' proficiency with number and number sense. They promote an efficient and flexible application of strategies to solve unknown problems.

Number Routines are a daily opportunity to focus on the development and strengthening of number sense and can be used throughout the day when time permits.

Number Routine
Would You Rather?

🕐 5–7 min

Build Fluency Students build skills with addition and estimation as they compare amounts.

These prompts encourage students to talk about their reasoning:

- What strategies did you use to find your answer?
- How can you use estimation to compare the number of pennies?

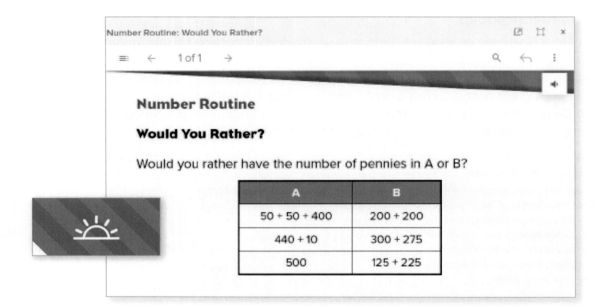

Learn more about teaching with Number Routines with the Professional Learning Resources in your Teacher Center.

44 Implementation Guide

Students revisit Number Routines across grades using the same structure but with more complex numbers or quantities. A full description of each routine can be found the Teacher Edition Appendix.

Routine	Grade K	Grade 1	Grade 2	Grade 3	Grade 4	Grade 5
About			✓	✓	✓	✓
Benchmark		✓	✓	✓	✓	✓
Break Apart	✓	✓	✓	✓	✓	✓
Can You Make It?					✓	✓
Counting Things	✓					
Greater Than Less Than		✓	✓	✓	✓	✓
Math Pics	✓	✓	✓	✓	✓	✓
Mystery Number			✓	✓		
Old and New Patterns	✓	✓	✓	✓	✓	✓
Pattern Count		✓	✓	✓		
Start and End	✓					
The Counting Path	✓					
The Match	✓					
The Missing			✓	✓	✓	✓
What Did You See?	✓				✓	
Would You Rather?	✓	✓	✓	✓	✓	✓
What's the Problem?		✓	✓	✓	✓	✓
Where Does It Go?		✓	✓	✓	✓	✓

Number Routines

Sense-Making Routines

Another set of routines in *Reveal Math* are **Sense-Making Routines**. Every lesson launches with the **Be Curious** activity that consists of a sense-making routine designed to build in students the habit of making sense of a situation, a foundational part of the problem-solving process.

When students spend time on just making sense of a problem situation — rather than trying to solve the problem — they begin to notice relationships and patterns that may not be as obvious and wonder what other relationships and patterns may exist or why these relationships and patterns exist.

Learn more about Sense-Making Routines with Professional Learning Resources in your Teacher Center.

Students will experience these four Sense-Making Routines throughout the *Reveal Math* program. These images are able to be shown to the class using the Lesson Presentations available in the Teacher Center.

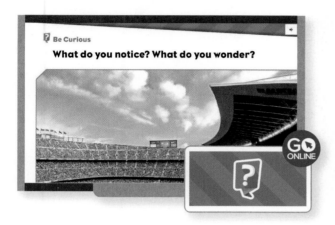

Notice and Wonder focuses students on making sense of the story, the quantities, and the real-world relationships of the mathematical concept.

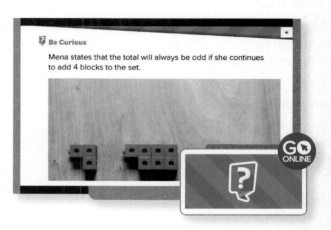

Is It Always True? presents students with images or situations that require thought about the relationship among the objects in the image. Students consider whether the relationship(s) is always true or unique to the image or situation.

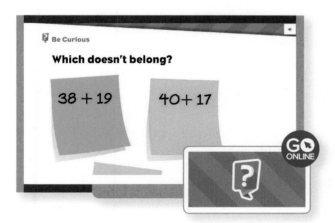

Which Doesn't Belong? presents a series of images, quantities, or numbers. Students compare and contrast the images or use reasoning to help identify which item "doesn't belong." The problem has multiple solutions depending on the reasoning students use.

Numberless Word Problems provide scaffolding that allows students the opportunity to develop a better understanding of the underlying structure of the problem itself.

Sense-Making Routines

Math Language Routines

Math Language Routines, developed by a team of authors at Center for Assessment, Learning, and Equity at Stanford University, are based on principles for the design of mathematics curricula that promote both content and language.

The team developed a framework with four design principles and eight Math Language Routines. Each routine aligns to one or more of the design principles.

The four design principles are:

Design Principle 1: Support sense-making

Design Principle 2: Optimize output

Design Principle 3: Cultivate conversation

Design Principle 4: Maximize meta-awareness

The eight Math Language Routines with their purposes are:

MLR1: *Stronger and Clearer Each Time* — Students revise and refine their ideas as well as their verbal or written outputs.

MLR2: *Collect and Display* — The teacher captures and displays students' words and phrases.

MLR3: *Critique, Correct, and Clarify* — Students analyze and then develop or revise a written argument.

MLR4: *Information Gap* — Students share unique pieces of information.

MLR5: *Co-Craft Questions and Problems* — Students generate questions to ask classmates.

MLR6: *Three Reads* — Students build comprehension skills as they read a passage three times, each time for a different purpose.

MLR7: *Compare and Connect* — Students compare and contrast different solution strategies, representations, or concepts.

MLR8: *Discussion Supports* — Students are provided prompts to support classroom discourse.

Reveal Math integrates Math Language Routines in every lesson during **Explore and Develop**.

Design Principle 1: Support sense-making

> **MLR Collect and Display**
> As you discuss the questions with the students, listen and write key words on the board that students use, such as *groups, objects, number of,* and *multiplication*. Display the words and phrases for student reference. Use the student-generated expressions to help them make connections between student language and math vocabulary. Update the collection with new understandings as the lesson progresses.

Design Principle 2: Optimize output

> **MLR Critique, Correct, and Clarify**
> On the board write, *There are 5 groups with 3 objects in each group*. Pair students to discuss whether this statement about the baskets of peaches is correct. Ask them to identify any mistakes and to make changes. Have students write a new, correct version of the sentence.

Design Principle 3: Cultivate conversation

> **MLR Stronger and Clearer Each Time**
> Have students justify why Greta does or does not have enough eggs. Have them first verbalize their ideas, and then share any information they know to justify it. Encourage students to question each other, clarify, and revise as needed.

Design Principle 4: Maximize meta-awareness

> **MLR Co-Craft Questions**
> Have students work alone to write questions about the problem and how to solve it. Then pair students to compare their questions. Elicit questions and use them to discuss the problem and strategies for solving it in more detail.

Fluency

Fluency is not about memorization or speed; it is much more complex and more appropriately refers to having efficient and flexible strategies for computing or solving equations.

Building fluency requires a working understanding and mastery of operations, relationships, and concepts.

Reveal Math provides students with multiple opportunities to revisit concepts and develop these areas of fluency within each unit. Number Routines provide students with daily opportunities to develop number sense, deepening their understanding of number relationships. In addition, every unit reviews a computational strategy previously learned to revisit concepts and strategies adding to students' flexibility when choosing methods.

We know fluency is not developed after one lesson, so *Reveal Math* provides ample opportunities for students to practice concepts. The **Game Station** provides opportunities to build on concepts from the lessons, while **Spiral Review** provides rotating review of previously learned concepts and skills.

Digital Station
The Digital Station includes games that offer an engaging environment to help students build computational fluency. The Digital Station is part of the Differentiated Support for each lesson.

Spiral Review
Spiral Review, available as a print-based or digital assignment, provides practice with mixed standard coverage for major clusters within the grade level to build fluency.

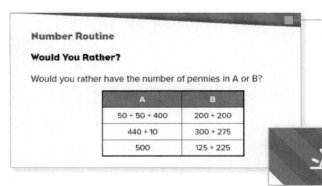

Number Routines help students develop strong number sense and an efficient and flexible application of strategies. They offer students daily opportunities to focus on number sense.

Unit Fluency Practice

Fluency Practice is also built into every Unit Review and includes:

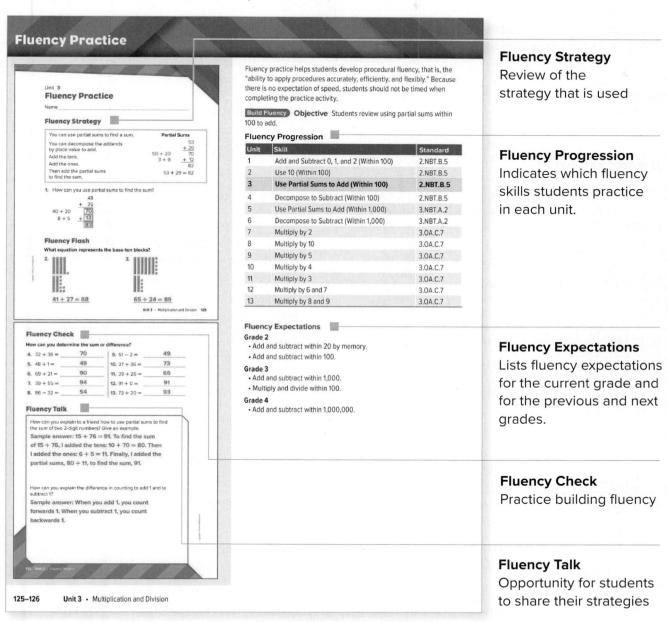

Fluency Strategy
Review of the strategy that is used

Fluency Progression
Indicates which fluency skills students practice in each unit.

Fluency Expectations
Lists fluency expectations for the current grade and for the previous and next grades.

Fluency Check
Practice building fluency

Fluency Talk
Opportunity for students to share their strategies

Assessments

Reveal Math offers a comprehensive set of assessment tools designed to be used in one of three ways:

- as a **diagnostic tool** to determine students' readiness to learn and diagnose gaps in students' readiness;
- as a **formative assessment tool** to inform instruction, and
- as a **summative assessment tool** to evaluate students' learning of taught concepts and skills.

The table below shows a listing of assessments in *Reveal Math*.

Assessments

TYPE	ASSESSMENT	HOW OFTEN	DESCRIPTION
DIAGNOSTIC	Course Diagnostic	Beginning of the school year	Diagnoses students' strengths and weaknesses with prerequisite concepts and skills for the upcoming year
	Unit Diagnostic	Beginning of each unit	Diagnoses students' strengths and weaknesses with prerequisite concepts and skills for the upcoming unit
FORMATIVE	Work Together	During a lesson	Assesses students' understanding of the concepts and skills presented in the Learn
	Exit Ticket	At the end of a lesson	Assesses students' conceptual understand and procedural fluency with lesson concepts and skills
	Math Probe	During a unit	Assesses common misconceptions
SUMMATIVE	Unit Assessment, Forms A and B	At the end of a unit	Evaluates students' understanding of and fluency with unit concepts and skills.
	Unit Performance Task	At the end of a unit	Evaluates students' ability to apply concepts and skills learned
	Benchmark Assessments	After multiple units	Evaluates students' understanding of concepts and skills taught in multiple units
	End of the Year Assessment	At the end of the school year	Evaluates students' proficiency with concepts and skills taught over the school year.

All assessments are available for either print or digital administration.

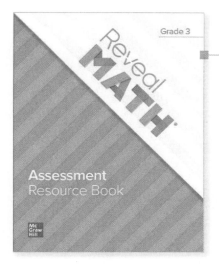

All print assessments are available as printable PDFs and are found in the Assessment Resource Book. Item analysis tables found in the Teacher Edition include recommendations for intervention support.

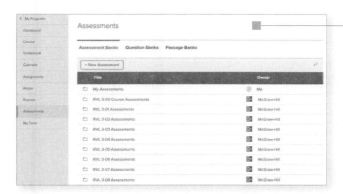

Digital assessments are customizable as needed. Teachers also have access to assessment item banks to build additional assessments as needed. Many of the digital assessment items are auto-scorable. Teachers have access to assessments reports in the Teacher Center. For more information on the digital reporting, see page 66.

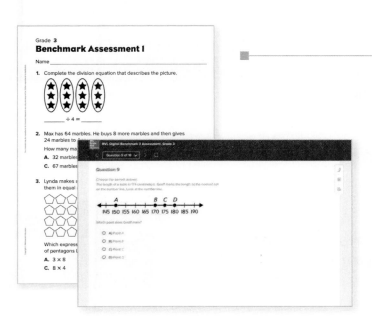

Reveal Math assessments include a range of item types that students are likely to encounter on high-stakes assessments. These include single-response multiple choice, multiple-response multiple choice, fill-in-the-blank, matching, and constructed response. The digital assessments include technology-enhanced items, such as drag and drop and drop-down menu select.

Math Probe – Formative Assessment

Target Common Misconceptions

Uncovering students' misconceptions is an integral part to teaching. *Reveal Math* offers structured opportunities to help teachers uncover common misconceptions related to key math concepts. In every unit is a **Math Probe** written by Cheryl Tobey designed to uncover students' misconceptions. These probes, placed at point of use, allow teachers to make sound instructional choices targeting specific mathematics concepts.

Short, targeted assessment

Each Math Probe has three to four 2-part items:

One part assesses students' understanding of concepts and

A second part asks students to share their thinking about the concepts.

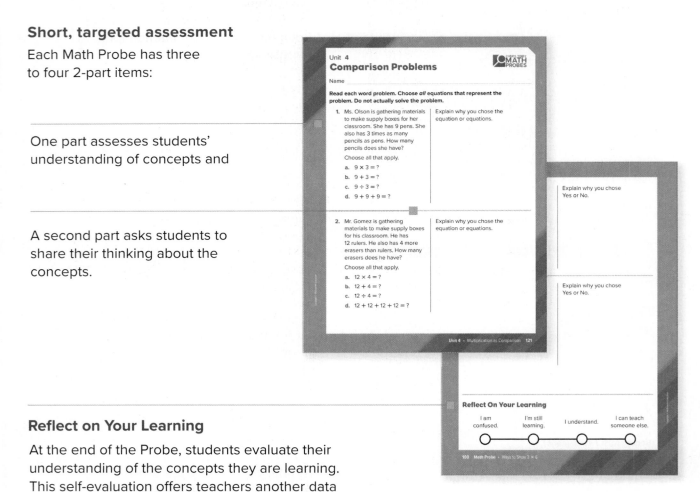

Reflect on Your Learning

At the end of the Probe, students evaluate their understanding of the concepts they are learning. This self-evaluation offers teachers another data point of students' understanding of the concepts.

Learn more about Math Probes with Professional Learning Resources in your Teacher Center.

54 Implementation Guide

Designed to ACT

The teacher support materials that accompany the Math Probes are designed around an ACT cycle — **Analyze** the Probe, **Collect** and Assess Student Work, and **Take** Action. The ACT cycle was originally developed during the creation of a set of math probes and teacher resources for a Mathematics and Science Partnership Project. The three teacher actions are:

A **Analyze the Probe**
Prior to administering the Math Probe, the teacher reviews the items and anticipates student difficulties. The targeted misconceptions are described in the Teacher Edition.

Analyze The Probe ✓ Formative Assessment

Targeted Concept Understand important multiplication ideas, such as "groups of," repeated addition, and skip counting. Recognize visual representations of multiplication, such as equal groups and arrays.

⚠ Targeted Misconceptions Students may focus on the product and select any representation based on that alone. They might not think about the value of the factors and the multiple of the operation. They may also not recognize that the first factor represents the number of groups; the second factor represents the size of each group.

Authentic Student Work
Below are examples of correct student work and explanations.

Sample A

C **Collect and Assess Student Work**
After administering the Math Probe, the teacher reviews students' responses and explanations to look for patterns of understandings and misunderstandings.

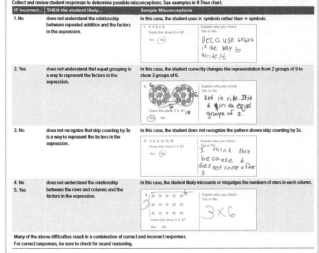

T **Take Action**
The teacher reviews and implements as appropriate the remedies provided that tie to specific misconceptions.

Math Probe — Formative Assessment

Differentiation Resources

Meeting the learning needs of all students is an important part of the *Reveal Math* vision. To meet these needs, *Reveal Math* includes a robust offering of differentiation resources for each lesson and unit. The variety of resources allows teachers to meet the learning needs of their students while also providing a range of implementation options, from independent to small group work completed in either a print or a digital environment.

Small Group Activities

Reinforce Understanding
These teacher-facilitated small group activities are designed to revisit lesson concepts for students who may need additional instruction.

How Many Xs?
Work with students in pairs. Have one student roll a number cube and then draw that number of circles. Then have the other student roll a number cube to determine the number of Xs to draw in each circle. Students should record a multiplication equation to find the total number of Xs. Help students recognize that they can skip count instead of counting all of the circles. Repeat the process. Have the students compare their totals over several rounds to determine the greatest number.

Build Proficiency
Students can work in pairs or small groups on the print-based Game Station activities, written by Dr. Nicki Newton, or they can opt to play a game in the Digital Station that helps build fluency.

Own it! Digital Station
Build Fluency Games
Assign the digital game to develop fluency with addition and subtraction.

Practice It! Game Station
Equal Groups Bingo
Students practice representing multiplication using equal groups.

Extend Thinking
The Application Station tasks offer non-routine problems for students to work on in pairs or small groups.

Use it! Application Station
How Many Beats in a Song? Students analyze sheet music to determine the number of beats in each measure of a song.

Independent Activities

Reinforce Understanding
Students in need of additional instruction on the lesson concepts can complete either the **Take Another Look** mini-lessons, which are digital activities, or the print-based **Reinforce Understanding** activity master.

Take Another Look Lesson
Assign the interactive lesson to reinforce targeted skills.
- Model Multiplication (Objects)

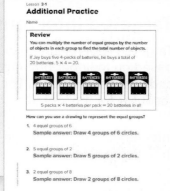

Build Proficiency
Additional Practice and **Spiral Review** assignments can be completed in either a print or digital environment. The digital assignments include learning aids that students can access as they work through the assignment. The digital assignments are also auto-scored to give students immediate feedback on their work.

Spiral Review
Assign the digital Spiral Review Practice to students or download and print PDFs of the Spiral Review from the Digital Teacher Center.

Interactive Additional Practice
Assign the digital version of the Student Practice Book.

Extend Thinking
The **STEM Adventures** and **Websketch** activities powered by Geometer's Sketchpad offer students opportunities to solve non-routine problems in a digital environment. The print-based **Extend Thinking** activity master offers an enrichment or extension activity.

Websketch Exploration
Assign a websketch exploration to apply skills and extend thinking.

STEM Adventures

Differentiation Resources **57**

Targeted Intervention

Reveal Math is committed to supporting all students to achieve high academic results. To that end, *Reveal Math* offers targeted intervention resources that provide additional instruction for students as needed.

Targeted Intervention at the Unit Level

Targeted intervention resources are available to assign students based on their performance on all Unit Readiness Diagnostics and Unit Assessments. The Item Analysis table lists the appropriate resource for the identified concept or skill gaps.

Intervention resources can be found in the Teacher Center in both the Unit Overview and Unit Review and Assess sections.

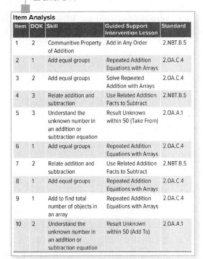

Item Analysis table from Teacher Edition

Teacher Center

Guided Support provides a teacher-facilitated small group mini-lesson that uses concrete modeling and discussion to build conceptual understanding.

Skills Support are skill-based practice sheets that offer targeted practice of previously taught items.

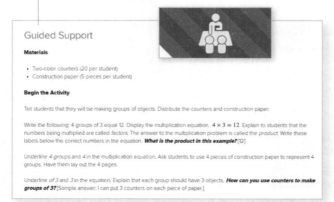

58 Implementation Guide

Targeted Intervention at the Lesson Level

Teachers can easily assign a **Take Another Look** mini-lesson for students to complete during independent work time, or they can be used in a small group to review a skill or concept. Each mini-lesson consists of a three-part, gradual-release activity that reteaches a key skill or concept. One to three Take Another Look lessons are identified for every lesson. These align to the end-of-unit assessment intervention resources.

Part One: Model Concept

A two- or three-minute video or animation introduces and models the skill or concept using essential math vocabulary.

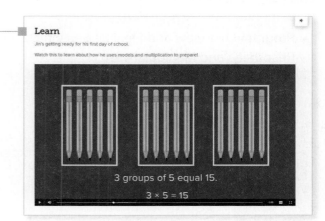

Part Two: Interactive Practice

A series of engaging activities provides students immediate feedback and helps build students' confidence through scaffolded repetition.

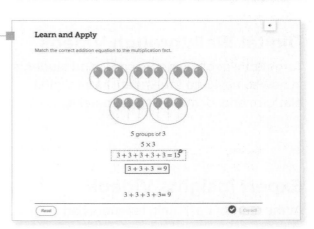

Part Three: Data Check

A quick three- to five-question assessment checks student understanding and provides teachers with data to inform instruction.

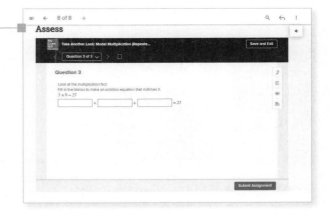

Professional Learning Resources

Reveal Math teachers have access to a comprehensive set of online professional learning resources to support a successful initial implementation and continued learning throughout the year. These self-paced, digital resources are available on-demand, 24 hours a day, 7 days a week in the Teacher Center for each grade.

Reveal Math Quick Start

The **Quick Start** includes focused, concise videos and PDFs that guide teachers step-by-step through implementing the *Reveal Math* program.

- Program Overview and Flexible Lesson Design
- Resources and Manipulatives
- Unit and Lesson Routines, Practices, and Processes
- Unit 1: Math Is...
- Digital Student Experience
- Social and Emotional Learning (SEL) Competencies
- Differentiation Pathways
- Assessments

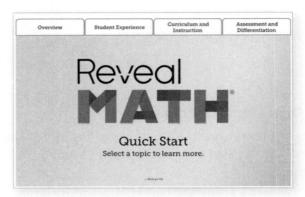

Digital Walkthrough Videos

Targeted videos guide teachers and students in how to navigate the *Reveal Math* digital platform and locate online resources.

Expert Insights Videos

At the start of each unit, teachers can view a 3-minute video of *Reveal Math* authors and experts sharing an overview of the concepts students will learn in the unit along with teaching tips and insights about how to implement the lesson.

Instructional Videos with *Reveal Math* Authors and Experts

- Annie Fetter: Be Curious Sense-Making Routines
- John SanGiovanni: Number Routines and Fluency
- Linda Gojak: Guided and Activity-Based Exploration
- Raj Shah: Ignite! Activities
- Cheryl Tobey: Math Probes
- Social and Emotional Learning

Model Lesson Videos

Classroom videos of *Reveal Math* lessons being taught to students show how to implement key elements of the Reveal Math instructional model.

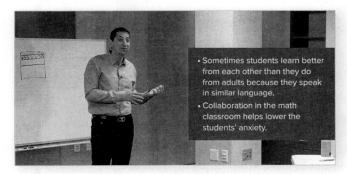

Ready-to-Teach Workshops

Curated, video-based learning modules on instructional topics key to *Reveal Math* can be used by teachers for self-paced learning or by district and school leaders as ready-to-teach packages to facilitate on-site or remote professional learning workshops.

Locating Your Resources

Professional Learning resources can be accessed in the drop-down menu under Course > Program Overview: Learning & Support Resources.

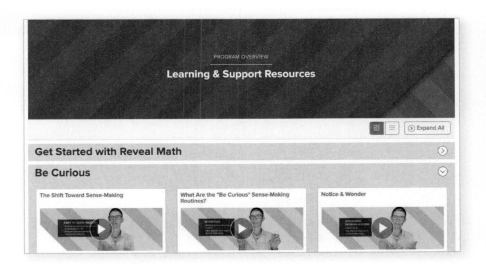

Professional Learning Resources **61**

Digital Experience

Student Center

The Student Dashboard is designed with our young learners in mind—allowing them to access all learning tools with ease.

Students can access specific lessons.

Students can review previously completed work and their scores on assignments.

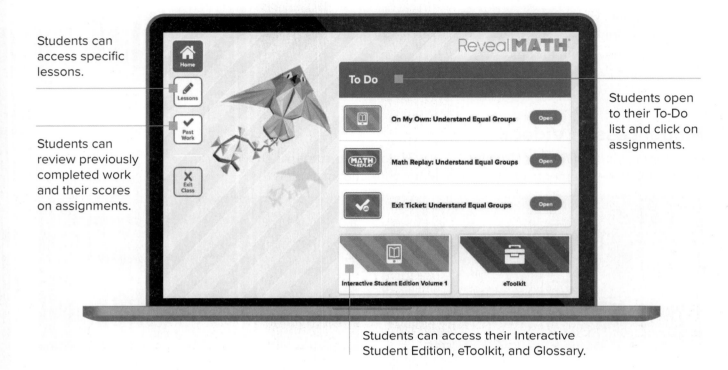

Students open to their To-Do list and click on assignments.

Students can access their Interactive Student Edition, eToolkit, and Glossary.

Interactive Student Edition

The Interactive Student Edition allows students to interact with the Student Edition as they would in print.

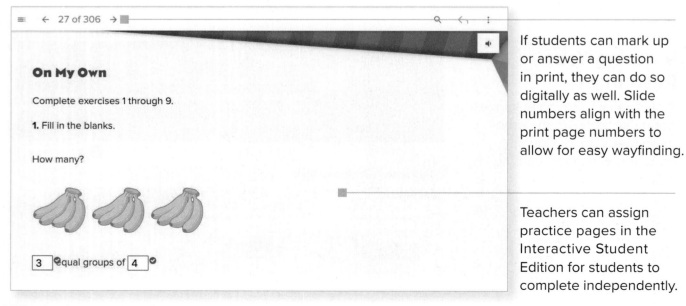

If students can mark up or answer a question in print, they can do so digitally as well. Slide numbers align with the print page numbers to allow for easy wayfinding.

Teachers can assign practice pages in the Interactive Student Edition for students to complete independently.

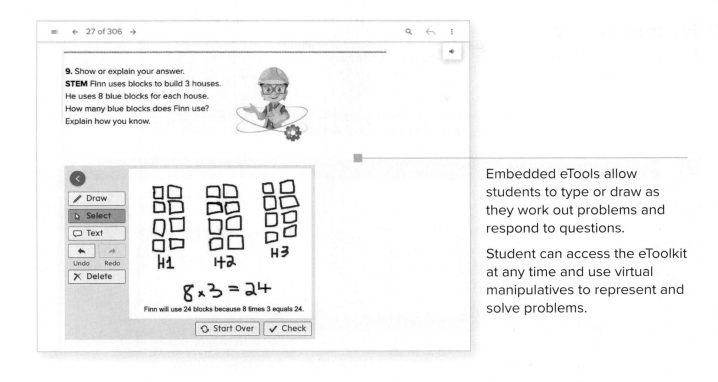

Embedded eTools allow students to type or draw as they work out problems and respond to questions.

Student can access the eToolkit at any time and use virtual manipulatives to represent and solve problems.

Digital Practice

Assigned Interactive Additional Practice and Spiral Review provide a dynamic experience, complete with learning aids integrated into items at point-of-use, that support students engaged in independent practice.

Digital Games

Digital Games encourage proficiency through a fun and engaging practice environment.

Digital Experience

Teacher Center

Teachers can access digital classroom resources and tools through the Teacher Center.

Browse the Course Navigation Menu to go directly to a unit or lesson.

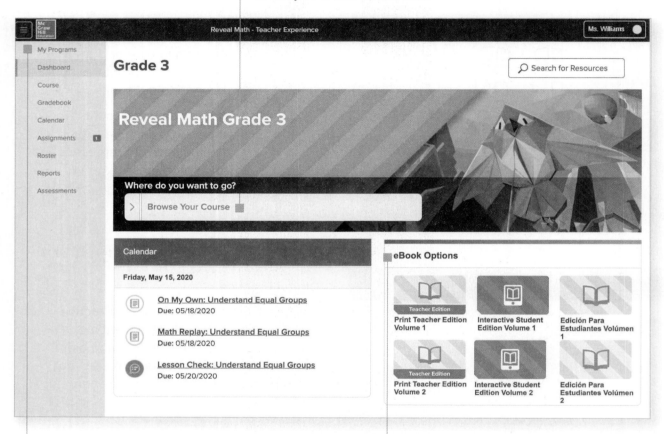

The menu provides access to:

- My Programs
- Dashboard
- Course View
- Table of Contents
- Gradebook
- Calendar
- Assignments
- Roster
- Reports
- Assessments

Shortcuts to the Interactive Student Edition and eBooks of the Teacher Edition and Spanish Student Edition are available on the dashboard.

Unit and Lesson Resource Pages

Unit and lesson resources are organized into landing pages for point-of-use access. Teachers can easily plan and prepare to teach units and lessons using the simple layout organization that aligns with their print Teacher's Edition.

Select a unit and lesson.

Launch/rearrange lesson presentations.

Add the lesson calendar for easy access.

Assign activities or assessments to a group, individual, or whole class.

See at a glance the different parts of the lesson.

Print lesson support resources at point-of-use.

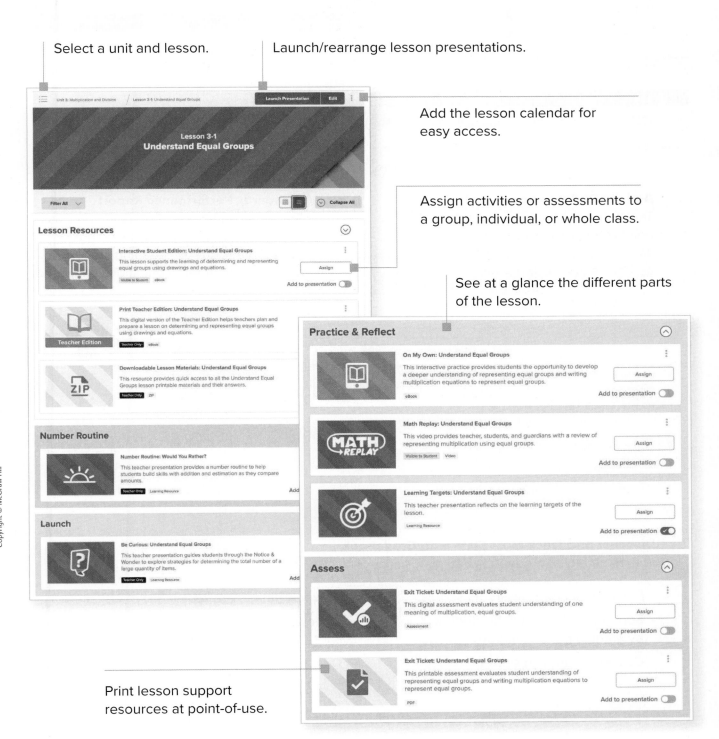

Digital Experience 65

Reporting

Interactive performance reports provide immediate feedback to teachers allowing them to make data-driven instructional decisions.

Activity Performance Report
Teachers can review useful data points for class activities, including item analysis by student and class, as well as overall performance.

Standards Performance Report
Teachers can access information on class performance by standard, including a cumulative score by class and student, as well as the number of questions answered.

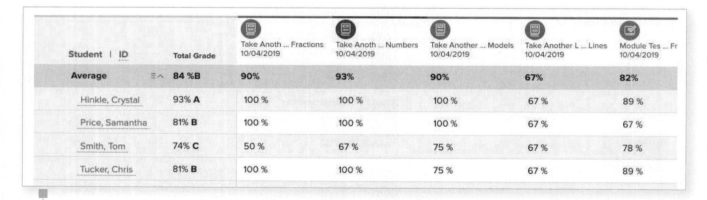

Discover and Track More Data with Gradebook
Within the digital gradebook, teachers can:

- Edit and manage classroom scores.
- Sort grades by group, by student, by grading period, and performance.
- Customize grading scales.
- Export data.
- View score sheets.

66 Implementation Guide

Class Management Tools

Class Management tools help maximize planning time.

Preview Student Experience
Emulate Student allows teachers to view which resources students will see and have access to their Student Digital Center. This will help ensure that assignments are set up correctly and will allow teachers to demonstrate usability and teach digital routines.

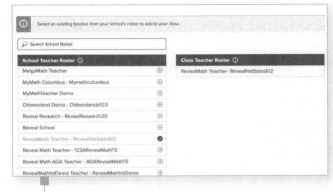

Share Your Class
Teachers can share class rosters, groupings, reports, assignments, lesson plans, and more with colleagues for the purpose of co-teaching, intervention, or instructional planning.

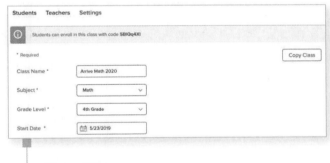

Copy Class
Copy functionality allows teachers to copy all assignments and customizations to another class.

Group Your Students
By making groups, teachers can differentiate assignments and assessments by instructional grouping as well as filter their gradebook by group.

Content Guide

STEM Career Kids
K-2 STEM Careers 70
3-5 STEM Careers 73

Social and Emotional Competencies
Self-Awareness 77
Self-Management 78
Social Awareness 79
Relationship Skills 80
Responsible Decision-Making 81

Key Concepts and Learning Objectives
Kindergarten 82
Grade 1 84
Grade 2 86
Grade 3 88
Grade 4 90
Grade 5 92

Scope and Sequence
Counting and Cardinality 94
Operations and Algebraic Thinking 96
Number and Operations in Base Ten 100
Number and Operations—Fractions 103
Measurement 105
Data 109
Geometry 110

Tables of Content & Pacing

Kindergarten 112
Grade 1 114
Grade 2 116
Grade 3 118
Grade 4 120
Grade 5 122

Correlations

Kindergarten 124
Grade 1 126
Grade 2 129
Grade 3 132
Grade 4 136
Grade 5 140

STEM Careers

Throughout *Reveal Math*, students are introduced to STEM careers through the STEM Career kids. Encourage students to learn more about these different careers and to see themselves in one (or more!) of these careers.

Grades K-2 STEM Career Kids

Emily, AEROSPACE ENGINEER

Aerospace engineers design aircraft, like airplanes or and missiles, or space craft like spaceships or satellites. They can also create prototypes of new aircraft or space craft and evaluate the performance of aircrafts. They are often called on to ssure all blueprints, prototypes, and products meet required engineering standards, environmental issues, and customer requirements.

Emily appears in Kindergarten, Unit 10, Grade 1, Unit 10, and Grade 2, Unit 6.

Jorden, ANIMAL TRAINER

Animal trainers work with different kinds of animals to get them to behave in a certain way and to respond to commands. Some animal trainers train service animals to help people with disabilities. Some trainers train animals to perform for an audience. They often work in zoos or aquariums or even on movie sets. Some trainers help families train their new pets, especially dogs. Animal trainers need a lot of patience and a love of animals.

Jordan appears in Kindergarten, Unit 7, Grade 1, Unit 5, and Grade 2, Unit 7.

Riley, AUTOMOTIVE ENGINEER

Automotive engineers design and develop cars, trucks, buses, motorcycles, and other off-road vehicles. They are also often called on to solve different engineering problems related to existing vehicles or prototypes. When designing new vehicles, automotive engineers must make sure the vehicle specifications align to all safety, energy, and environmental requirements.

Riley appears in Kindergarten, Unit 2, Grade 1, Unit 6, and Grade 2, Unit 9.

Chloe, CARPENTER

Carpenters are skilled workers who work with wood constructions. They can build houses, boats, pieces of furniture, kitchens, or make molding and trim for a building or house. Carpenters work with a lot of tools, such as saws, hammers, and tape measures. Carpenters use math a lot to measure lengths, heights, and widths.

Emily appears in Kindergarten, Unit 14, Grade 1, Unit 7, and Grade 2, Unit 12.

Kayla, LANDSCAPE ARCHITECT

Landscape architects plan and design public outdoor spaces, such as parks, gardens, cemeteries, recreational spaces, college or school campuses, and other open spaces. Their job is to enhance the natural beauty of a space and to foster environmental benefits. They sometimes plan the restoration of a natural environment that has been changed by humans, such as wetlands.

Kayla appears in Kindergarten, Unit 11, Grade 1, Unit 13, and Grade 2, Unit 10.

Hugo, METEOROLOGIST

Meterologists study the Earth's atmosphere and the environment. Many meterorlogists report on the weather and weather conditions on TV or radio, but others study ways to reduce air pollution and global warming. Some work for government agencies, such as NASA or the National Weather Service.

Hugo appears in Kindergarten, Unit 9, Grade 1, Unit 11, and Grade 2, Unit 11.

Sienna, NUTRITIONIST

Nutritionists work in the field of food and nutrition. With their expertise in nutrition, they advise people on what to eat to lead a healthy lifestyle or achieve a specific health-related goal. They may develop meal plans for people who need to eat a specific diet or who want to achieve a certain life goal, monitor the effects of the diet, and make adjustment to the diet as needed. Nutritionists work in many settings, including hospitals, cafeterias, nursing homes, and schools.

Sienna appears in Kindergarten, Unit 13 and Grade 2, Unit 2.

Jin, PALEONTOLOGIST

Paleontologist are scientists who study the fossilized remains of plants and animals. They study fossils to learn about extinct plants and animals, the history of life on earth, and most importantly, to draw inferences about what might happen to life on earth in the future. Paleontologists often work with archeologists at dig sites.

Jin appears in Kindergarten, Unit 4 and Grade 1, Unit 3.

Marisol, PARAMEDIC/ EMT

Paramedics and emergency medical technicians (EMTs) take care of the medical needs for sick or injured people. They often work for city or town fire departments or for ambulance companies. Being a paramedic can be physically demanding if they are treating adults. Paramedics often need to work very quickly and make sometimes life-saving decisions quickly. They often comfort and calm their patients

Marisol appears in Kindergarten, Unit 6, Grade 1, Unit 9, and Grade 2, Unit 3.

Deven, SOUND ENGINEER

Sound engineers love to make new sounds by blending different sounds. They often work in recording studios to make sure a singer's voice blends well with the instruments. They can also work with live concerts in outdoor arenas, concert halls, or even small restaurants. Sound engineers understand electronics and acoustics.

Deven appears in Kindergarten, Unit 8, Grade 1, Unit 12, and Grade 2, Unit 4.

C.J., STATISTICIAN

Statisticians collect data and then use different mathematical techniques to analyze and interpret data, and draw conclusions based on the data. Statisticians often participate in designing a survey or experiment. Sometimes they analyze data that others have collected.

C.J. appears in Kindergarten, Unit 3, Grade 1, Unit 4, and Grade 2, Unit 8.

Erik, VIDEO GAME DESIGNER

Who doesn't love a great video game? Video game designers design video games, from pinball-type games to complete immersive experiences. Video game designers write many many lines of code to create all of characters in a game and to make them move around. Writing code requires a lot of attention to detail.

Erik appears in Kindergarten, Unit 13, Grade 1, Unit 2, and Grade 2, Unit 5.

Grades 3-5 STEM Career Kids

Sam, ARCHITECT

Architects design places where people live, work, learn, shop, worship, and meet. These places can be private, like residences, or public, like schools or shopping centers. Architects create the look of a building and are also responsible for ensuring the building meets a variety of requirements from safety to functionality. Architects are involved in nearly all parts of the construction of the buildings they have designed. Architects often specialize in a kind of building, such as educational institutions, office buildings, or residential buildings.

Sam appears in Grade 3, Units 6 & 11 and Grade 5, Unit 13.

Haley, ASTRONOMER

Astronomers are scientists who study space, including the stars, the planets, and the galaxies. They are the oldest of scientists, with evidence of the ancient Greeks studying the stars as early as the 4th century BC. Astronomers study the life cycle of stars and planets to gain an understanding of how they were created and what will happen to them as they age. Astonomers still use telescopes to study the planets, but modern technology has allowed for more precise measurments and information.

Haley appears in Grade 3, Unit 7, Grade 4, Unit 5, and Grade 5, Unit 3.

Saffron, CHEF

Chefs work in restaurants or other places where food is prepared. In restaurants, chefs plan the menu, maintain the budget, purchase supplies, manage the team that prepares and cooks the meals, ensuring the quality of food being served. Chefs need to have a comprehensive understanding of food preparation and processing and food safety.

Saffron appears in Grade 3, Unit 2, Grade 4, Unit 11, and Grade 5, Unit 8.

Grace, COMPUTER PROGRAMMER

Computer programmers create software programs by writing code. They test the programs they have created to make sure they work correctly. A big part of the work of computer programmers is finding and correcting bugs in the code when necessary. Programmers often write thousand and thousands of lines of code for a software program.

Grace appears in Grade 3, Unit 4, Grade 4, Unit 12, and Grade 5, Unit 7.

Finn, CONSTRUCTION MANAGER

Construction managers supervise a wide variety of building projects, from residential to commercial and industrial structures, such as roads, bridges, schools, and hospitals. Construction managers are responsible for cost estimates, budgets, and schedules. They oversee the hiring of workers to work on different parts of the project. They often problem solve to address delays and other work-related problems.

Finn appears in Grade 3, Unit 3, Grade 4, Unit 7, and Grade 5, Unit 12.

Owen, ENTOMOLOGIST

Entomologists are scientists who study insects. They study the life cycle and habits of insects. Entomologists usually specialize in a specific category of insects, such as bees, butterflies, or beetles. Entomologists study insects and their relationship to humans and the environment. They may specialize in agriculture, ecology, or insecticide toxicology.

Owen appears in Grade 3, Unit 8, Grade 4, Unit 9, and Grade 5, Unit 5.

Maya, GEOLOGIST

Geologists study earth processes such as earthquakes, landslides, floods, and volcanic eruptions. They study these processes to understand them better and to be able to predict when a major event might occur. Some geologists investigate earth materials, such as oil and natural gas, and develop and refine methods to extract these materials. Geologists often specialize in an area of geology, such as environmental geology, petroleum geology, or engineering geology.

Maya appears in Grade 3, Unit 12, Grade 4, Unit 6, and Grade 5, Unit 6.

Noah, NURSE

Nurses are healthcare professionals who oversee the day-to-day care of patients. Nurses who work in hospitals have a variety of responsibilities from administering medication, helping patients prepare for surgery, and assisting doctors and surgeons during procedures.

Noah appears in Grade 3, Unit 5 and Grade 4, Unit 13.

Hiro, OCEAN ENGINEER

Ocean engineers design and build equipment that can operate in the ocean environment. The equipment they design and build has to withstand the wear and tear of frequent use and the harsh conditions of the ocean environment. Ocean engineers may work in the oil and gas industry or in marine navigation sectors.

Hiro appears in Grade 3, Unit 10, Grade 4, Unit 3, and Grade 5, Unit 2.

Poppy, PARK RANGER

Park rangers work to maintain natural landscapes. Park rangers can also be called conservationists, environmentalists, or foresters, all of whom work tirelessly to protect the world's natural resources. Park rangers often work for state, regional, or federal agencies. Some park rangers work for private landowners to provide recommendations on conservation efforts. One of the most important duties of park rangers is to help prevent wildfires.

Poppy appears in Grade 4, Unit 2, and Grade 5, Unit 9.

Malik, PHOTONICS ENGINEER

Photonics engineers develop innovative products that use photonics, that is, light, to generate energy, detect information, or transmit information. Photonics engineers develop products for a variety of uses, from medicine to telecommunications and manufacturing. Photonics engineers often work with lasers for precision work, such as delicate surgeries or cutting precious metals.

Malik appears in Grade 3, Unit 9, Grade 4, Unit 8, and Grade 5, Unit 14.

STEM Careers

Antonio, ROBOTIC ENGINEER

Robotics engineers are responsible for creating robots and robotic systems that are able to perform duties that humans are either unable or prefer not to complete. Through their creations, robotics engineers help to make jobs safer, easier, and more efficient, particularly in the manufacturing industry.

Antonio appears in Grade 4, Unit 14 and Grade 5, Unit 11.

Ruby, VETERINARIAN

Veterinarians are medical professionals who treat animals, both large and small. During their medical training, veterinarians can choose to work with large animals, like cows and horses, or small animals, like cats and dogs. Just as doctors swear to protect the well-being of their human patients, veterinarians take a solemn oath to protect the well-being of the animals that they treat. Veterinarians need strong analytical skills since their patients usually cannot tell them how they are feeling or what their symptoms are.

Ruby appears in Kindergarten, Unit 5, Grade 1, Unit 8, Grade 4, Unit 10, and Grade 5, Unit 4.

Hannah, WELDER

Welder are skilled trades people who join metal together using heat. Welders work on all types of metals in all types of construction applications. Some may even work underwater to repair oil rig foundations, ship hulls, and other types of subaquatic structures.

Hannah appears in Grade 3, Unit 13, Grade 4, Unit 4, and Grade 5, Unit 10.

Social and Emotional Learning Competencies

Reveal Math developed its SEL support from the CASEL framework, which consists of five competencies.

Self-Awareness

Self-awareness is the ability to identify and describe one's emotions, thoughts, and needs and to understand how they influence behavior. It includes the ability to assess one's strengths and limitations, with a well-grounded sense of confidence and a "growth mindset." When students, especially young students, have an understanding of their emotions, thinking, and needs, they are more likely to achieve success in school and life. Self-awareness is also critical for recognizing emotions in others.

Self-Awareness Sub-competency	Grade K	Grade 1	Grade 2	Grade 3	Grade 4	Grade 5
Accurate Self-Perception Students are encouraged to identify areas of strength and of growth related not just to math, but to other academic areas as well as non-academic areas, to build an accurate self-perception.	3-6 7-2 9-5	6-2 9-8 13-1	5-2 7-2 9-4	2-10 4-1 9-6	7-3 9-5 13-6	5-1 11-6 13-4
Identify emotion Helping students recognize and identify emotions helps them become better equipped with managing emotions, especially negative emotions that could impact their ability to focus on their learning.	8-3 12-4 13-5	3-1 4-5 8-5 11-1	3-7 6-5 7-5 10-9	2-1 6-5 8-3 11-2	3-3 4-1 8-4 11-4	4-2 5-7 8-4, 10-8
Recognize strengths When students recognize their strengths, they can be encouraged to use their strengths in different ways to help achieve academic success.	5-1 8-8 10-6 11-2	2-3 3-2 7-3 9-3 12-3	2-2 4-10 6-4 8-2	2-6 5-1 7-3 9-8 12-2	5-1 10-5 14-4	4-7 7-1 9-1 11-2 14-4
Self-confidence Building students' self-confidence can help them address challenges and problems they may encounter in their studies.	2-1 3-3 10-2 14-5	4-1 6-4 10-4	4-6 6-9 11-5	5-5 10-3 12-10	2-3 3-4 9-1 12-4	3-1 6-3 10-1
Self-efficacy Self-efficacy relates to students' belief in their ability to succeed in a given situation, specifically in a school setting and the math classroom.	2-7 4-4 6-1	5-3 8-3 12-2	5-7 10-2 12-4	3-5 8-7 10-6	6-5 13-1 14-5	9-9 12-1

Self-Management

Self-management requires skills and attitudes that help students to regulate their emotions and behaviors. This includes the ability to delay gratification, manage stress, control impulses, and persevere with tasks, especially challenging ones, to achieve personal and educational goals. Research has shown that preschool children can begin to self-manage as they remember and generalize appropriate behaviors and monitor their behaviors in different situations.

	Grade K	Grade 1	Grade 2	Grade 3	Grade 4	Grade 5
Control impulses Controlling impulses is an important part of school readiness; at the same time, students often struggle with controlling impulses. When students have strategies to build better impulse control, they can be more successful staying focused.	2-3 5-3 10-3	4-10 6-3 12-1 12-10	3-5 7-3 10-1 11-3	2-4 5-6 13-1	7-1 10-2 12-2	2-3 5-2 9-4
Goal setting Getting students thinking about setting goals helps them build confidence and focus while also improving their decision-making.	2-5 8-2 11-4	3-3 9-6	4-9 7-10 8-1	2-11 10-1 12-4	3-6 6-2 9-3	4-1 7-6 10-9
Manage stress It's important to help students build coping strategies to manage stress and negative feelings. As students develop these strategies, they are more likely to experience academic success.	3-11 6-5 7-1	3-7 4-7 13-2	4-5 6-3 9-1	6-2 9-5 12-1	3-9 6-4 7-7 14-2	3-2 6-4 11-1
Organizational skills Developing strong organizational skills will help students throughout their academic career, as well as in their personal, and later, professional lives.	3-8 8-6 13-4	7-1 10-2 12-4	4-7 5-6 9-5	2-5 3-6 8-5	5-5 8-3 13-11	10-5 11-4 13-3
Self-discipline As students learn to control their impulses and manage stressful situations, they learn to be self-disciplined so they can focus on learning.	3-5 9-2 12-3	2-1 5-7 11-4	5-4 7-7 12-5	4-2 9-3 11-5	4-3 10-6 14-8	5-5 8-2 14-3
Self-motivation Helping students be self-motivated shifts the motivation to learn from an extrinsic goal to an intrinsic one.	3-2 14-4	4-3 5-9 8-1	6-8 10-5	3-2 5-2 7-2 12-8	2-1 11-5 13-7 13-9	7-2 8-6 12-2

Social Awareness

Social awareness involves the ability to understand, empathize, and feel compassion for others, especially for those from different backgrounds or cultures. It also involves understanding social norms for behavior and recognizing family, school, and community resources and supports.

	Grade K	Grade 1	Grade 2	Grade 3	Grade 4	Grade 5
Appreciate Diversity It is important for students to understand that people come from a range of ethnic, cultural, and social backgrounds and have a wide range of abilities and interests.	3-7 4-3 9-6 10-1	5-4 6-5 8-6 13-3	3-6 5-8 6-10 8-3 11-6	3-3 5-3 7-1 11-1	3-1 6-6 8-5 12-1	2-2 6-6 9-5 12-5
Develop perspective Helping students appreciate the diversity of people they are likely to encounter, and understand the value of different viewpoints and perspectives will open students' minds to creative problem solving.	3-4 6-4 7-4 12-5 13-3	3-5 7-5 10-3 11-3	4-2 6-1 9-3 12-6	4-4 6-6 8-2 13-3	2-4 7-5 11-3 14-1	3-4 5-4 10-6 13-5
Empathy Empathy is the ability to understand and feel what another person is experiencing. Empathy is different from sympathy in that one is able to share the feelings of the other.	2-6 5-4 11-5 14-2	4-11 8-2 9-4 12-6	5-3 7-1 9-6	2-3 2-9 9-7 12-3	3-7 6-3 10-4 13-5 14-6	4-4 6-1 8-5 14-2
Respect Others When students respect one another they accept others as they are, for who they are, in spite of differences of ideas, viewpoints, or abilities. Respecting others creates mutual feelings of respect and trust, safety, and well-being.	2-4 8-4 9-1 11-3	2-5 4-4 5-8 12-5	4-8 7-8 10-4 12-1	6-4 9-2 10-5 12-7	5-3 7-2 9-2 13-3	4-8 7-5 9-2 11-3 14-5

Relationship Skills

Relationship skills build off students' social awareness and help them establish and maintain healthy and rewarding relationships, and to act in accordance with social norms. These skills involve communicating clearly, listening actively, cooperating, resisting inappropriate social pressure, negotiating conflict constructively, and seeking help when it is needed.

Relationship Skills	Grade K	Grade 1	Grade 2	Grade 3	Grade 4	Grade 5
Build Relationships Children who are able to build relationships with peers experience higher levels of emotional well-being. An important part of the school setting is helping students begin to build relationships not just with peers, but with adults other than their family members.	7-3 8-5 10-5	5-6 6-6 8-7 12-7	5-1 8-5 11-2 12-3	4-5 6-1 8-1 10-2	3-2 11-1 13-4 14-10	2-4 6-2 8-3 10-7
Communication Strong communication skills help students convey their thinking and emotions and to understand the thinking and emotions of others. Communication skills include not just speaking, but listening as well.	2-2 5-2 8-7 12-2	2-4 4-2 5-2 9-7 11-5 13-4	6-2 9-7 11-4	3-1 5-4 9-4 13-2	2-2 7-4 8-1 13-10	3-3 9-3 11-7 14-1
Social Engagement Social engagement relates to participation in activities with peers. To do so effectively, students need to learn and practice important social skills.	3-9 4-1 13-2 14-3	3-4 4-8 9-1 10-1	4-4 6-6 7-11 10-8	2-2 7-4 8-4 12-5	3-5 6-1 9-4 12-3	4-3 5-6 10-3 12-4
Teamwork Teamwork requires collaboration, mutual respect, and individual accountability. It requires compromise and a willingness to put group goals over individual wants.	3-1 6-5 9-4 11-6 13-6	3-6 7-6 12-8	5-9 7-4 10-3	2-8 7-6 11-3 12-11	4-2 5-2 6-7 10-3 14-3	4-5 7-3 9-7 13-1

Responsible Decision Making

Responsible decision making involves learning how to make constructive choices about personal behavior and social interactions in a range of settings. It requires the ability to consider ethical standards, safety concerns, accurate behavioral norms for risky behaviors, the health and well-being of self and others, and to make realistic assessments of consequences.

Responsible Decision-Making Skills	Grade K	Grade 1	Grade 2	Grade 3	Grade 4	Grade 5
Analyze situations A first step in responsible decision-making is understanding the problem situation. That includes analyzing the variables of the situation.	5-5 13-1	7-4 12-9	5-10 11-1	3-7 9-1 11-4	6-8 10-1 12-5	5-3 8-1 12-3
Identify problems The second step in responsible decision-making is to identify the problem to be solved. Problems include not just academic problems, but interpersonal problems as well.	6-2 7-5 11-1	2-2 7-2 8-4	2-4 4-1 9-2	5-7 7-5 13-4	3-8 11-2 14-7	4-6 7-7 9-6
Solve problems Students need to develop problem-solving skills not just for use in the math classroom, but for use in their lives.	2-8 8-1 9-3	4-9 5-5 9-5	4-3 6-7 10-6	2-7 4-6 12-9	4-4 7-8 8-2	2-5 9-8 10-4
Evaluate Evaluating a process is useful in determining what works well and what could be improved in a program or initiative. Instilling in students the habit of evaluating decisions builds their decision-making competency.	2-9 3-10 10-4	5-1 13-5	3-1 10-7	2-12 6-3 10-4	5-6 13-2	6-5 11-5 13-6
Reflect Another aspect of responsible decision-making is reflection. Reflection encourages examination of oneself, one's behaviors, actions, and decisions.	3-12 14-1	3-8 9-2	5-5 7-9 12-2	4-3 8-6	5-4 14-9	3-5 10-2 13-2
Ethical Responsibility Ethical responsibility relates to the expectation or even duty to act in a way that is consistent with classroom values.	4-2 12-1	6-1 11-2	7-6 8-4	3-4 9-9 12-6	7-6 9-6 13-8	7-4 14-6

Key Concepts and Learning Objectives

Kindergarten

Habits of Mind and Classroom Norms

Students

- describe ways they use math in their lives and their world. (Unit 1)
- describe their strengths as doers of math. (Unit 1)
- explain what a problem is. (Unit 1)
- see mathematics in the real world. (Unit 1)
- explain their thinking. (Unit 1)
- work well on their own and in a group. (Unit 1)
- know the steps to take to solve a problem. (Unit 1)
- describe patterns. (Unit 1)

Representing and Comparing Whole Numbers

Students

- count objects to 10. (Units 2, 3)
- show numbers 1–10. (Units 2, 3)
- identify 0. (Unit 2)
- explain how to identify the number that is one more. (Units 2, 3)
- tell whether groups are equal. (Units 2, 3)
- compare two groups of objects or numbers. (Units 2, 3)
- write numbers to show how many. (Unit 3)
- represent numbers 11–19. (Units 9, 10)
- make groups of 11–19 objects. (Units 9, 10)
- decompose groups of 11–19 objects. (Units 9, 10)
- count by 1s and 10s to 100. (Unit 12)
- describe patterns when counting by 1s and 10s to 100. (Unit 12)
- count by 1s to 100, starting at any number. (Unit 12)
- count to answer "how many?" about as many as 20 things. (Unit 12)

Addition and Subtraction

Students

- represent and solve addition and subtraction problems. (Units 6, 7)
- solve addition and subtraction equations within 5 fluently. (Unit 8)
- compose and decompose numbers to 10 in different ways. (Unit 8)

Measurement and Data

Students

- sort and describe objects by attribute. (Unit 4)
- describe and compare objects using length, height, weight, and capacity. (Unit 14)

Describing Shapes in Space

Students

- name and describe 2-dimensional shapes. (Unit 5)
- describe the relative position of 2-dimensional and 3-dimensional shapes. (Units 5, 11)
- name and describe 3-dimensional shapes. (Unit 11)
- compare and contrast 2-dimensional and 3-dimensional shapes. (Unit 13)
- draw and build 2-dimensional and 3-dimensional shapes. (Unit 13)

Grade 1

Habits of Mind and Classroom Norms for Productive Math Learning

Students

- make sense of problems and think about numbers and quantities. (Unit 1)
- share their thinking with their classmates. (Unit 1)
- use math to make sense of everyday problems. (Unit 1)
- see patterns in math. (Unit 1)
- describe their math stories. (Unit 1)
- work productively with their classmates. (Unit 1)

Addition and Subtraction

Students

- relate counting to addition. (Unit 4)
- relate counting to and counting back to subtraction. (Unit 5)
- use different strategies to add and subtract within 100. (Units 4, 5, 9, 11)
- use addition to solve problems involving adding to and putting together. (Unit 7)
- use subtraction to solve problems involving taking from and taking apart. (Unit 8)
- use addition and subtraction to represent and solve compare problems. (Unit 10)
- explain what the equal sign means. (Unit 4)

Number Sense and Place Value

Students

- read and write numbers from 0 to 120. (Unit 2)
- use place value to represent 2-digit numbers. (Unit 3)
- explain that 10 ones equal 1 ten. (Unit 3)
- compare two 2-digit numbers by comparing the number of tens and the number of ones. (Unit 3)

Measurement

Students

- order three objects from shortest to longest. (Unit 12)
- compare the lengths of two objects. (Unit 12)
- measure the length of objects. (Unit 12)

Data

Students

- tell time to the nearest hour and half hour. (Unit 12)
- organize and interpret data into three categories. (Unit 12)

Attributes of Shapes

Students

- describe attributes that define shapes. (Unit 6)
- describe attributes that do not define shapes. (Unit 6)
- compose two-dimensional and three-dimensional figures. (Unit 6)

Grade 2

Habits of Mind and Classroom Norms

Students

- make sense of problems and think about numbers and quantities. (Unit 1)
- share thinking with classmates. (Unit 1)
- represent problems with mathematics. (Unit 1)
- use patterns to solve problems. (Unit 1)
- describe their math stories. (Unit 1)
- work well with classmates. (Unit 1)

Addition and Subtraction

Students

- write equations to describe arrays. (Unit 3)
- represent and solve one- and two-step word problems using addition and subtraction strategies. (Units 4, 5, 6, 9, 10)
- add addends in any order to find the sum. (Unit 5)
- add and subtract fluently within 20. (Units 5, 6)
- use tools to add and subtract. (Units 5, 6)
- add and subtract 2-digit and 3-digit numbers with and without regrouping. (Units 5, 6, 9, 10)
- mentally add 10 and 100 to a 3-digit number and subtract 10 and 100 from a 3-digit number. (Units 9, 10)
- explain how to use strategies to add and subtract 3-digit numbers. (Units 9, 10)

Whole Numbers

Students

- identify the digits in a 3-digit number. (Unit 2)
- read and write numbers to 1,000. (Unit 2)
- decompose 3-digit numbers in different ways. (Unit 2)
- compare 3-digit numbers. (Unit 2)
- identify and describe patterns when counting by 1s, 5s, 10s, and 100s. (Unit 3)
- determine the value of a group of coins. (Unit 8)
- tell time using analog and digital clocks. (Unit 8)

Measurement

Students

- measure and compare lengths using customary and metric units. (Unit 7)
- use everyday items to help estimate length in customary and metric units. (Unit 7)
- solve problems involving length. (Unit 7)

Data

Students

- collect measurement data. (Unit 11)
- interpret data on a line plot. (Unit 11)
- make a line plot to show data. (Unit 11)

Describe and Analyze Shapes

Students

- describe 2-dimensional and 3-dimensional shapes. (Unit 12)
- identify equal shares. (Unit 12)
- partition 2-dimensional shapes into equal shares. (Unit 12)
- partition rectangles into rows and columns of equal-sized squares. (Unit 12)

Grade 3

Habits of Mind and Classroom Norms for Productive Math Learning

Students

- make sense of problems and think about numbers and quantities. (Unit 1)
- share thinking with classmates. (Unit 1)
- use math to represent everyday problems. (Unit 1)
- see patterns in math. (Unit 1)
- describe their math stories. (Unit 1)
- work productively with classmates. (Unit 1)

Add and Subtract within 1,000

Students

- use different strategies to add 3-digit numbers. (Unit 2)
- use different strategies to subtract 3-digit numbers. (Unit 2)

Understand Multiplication and Division

Students

- use equal groups and arrays to represent multiplication. (Unit 3)
- show that the order of two factors in a multiplication equation does not change the product. (Unit 3)
- represent division with equal sharing and equal grouping. (Unit 3)
- describe and use patterns to multiply by 0, 1, 2, 5, and 10. (Unit 4)
- decompose a factor to multiply. (Unit 5)
- use properties of multiplication to recall multiplication facts. (Unit 5)
- use related multiplication facts to divide by numbers 1 through 10. (Unit 9)
- group three factors in different ways to multiply and explain how grouping factors can make it easier to multiply three numbers. (Unit 10)

Understand Fractions

Students

- explain the meanings of the numerator and the denominator of a fraction. (Unit 7)
- identify and represent fractions on a number line. (Unit 7)
- determine whether two fractions are equivalent. (Unit 8)
- generate equivalent fractions and explain why fractions are equivalent. (Unit 8)
- compare fractions with the same denominator and different numerators or fractions with the same numerator and different denominators. (Unit 8)

Understand Area

Students

- demonstrate understanding of concepts of area measurement. (Unit 6)
- determine the area of a rectangle using multiplication. (Unit 6)
- determine the area of composite figures. (Unit 6)
- solve problems involving area and perimeter. (Unit 11)

Measurement

Students

- measure objects to the nearest half and quarter inch. (Unit 12)
- solve word problems involving liquid volume and mass. (Unit 12)

Data

Students

- analyze information presented in scaled picture graphs and scaled bar graphs to solve problems. (Unit 12)
- create line plots to display measurement data. (Unit 12)

Describe and Analyze Two-Dimensional Shapes

Students

- describe polygons and classify them based on their shared attributes. (Unit 13)
- identify and classify quadrilaterals based on their attributes. (Unit 13)
- use given attributes and an understanding of categories of quadrilaterals to draw quadrilaterals. (Unit 13)

Grade 4

Habits of Mind and Classroom Norms for Productive Math Learning

Students

- make sense of problems and represent numbers and quantities in different ways. (Unit 1)
- share their thinking with classmates. (Unit 1)
- represent a real-world situation using mathematics. (Unit 1)
- use patterns to develop efficient strategies to solve problems. (Unit 1)
- describe their math biographies. (Unit 1)
- work productively with classmates. (Unit 1)

Place Value, Multi-Digit Arithmetic, and Properties of Operations

Students

- read and write numbers up to one million in multiple forms. (Unit 2)
- round multi-digit numbers to any place-value position. (Unit 2)
- add and subtract whole numbers within 1,000,000 using the standard algorithm. (Unit 3)
- solve multi-step word problems using the four operations, and assess the reasonableness of answers. (Units 3, 4, 6, 7)
- distinguish multiplicative comparison from additive comparison. (Unit 4)
- generate a number or shape pattern that follows a given rule, and identify apparent features of the pattern that were not explicit in the rule itself. (Unit 5)
- find all factor pairs for a whole number from 1–100. (Unit 5)
- determine whether a given whole number in the range 1–100 is a multiple of a given one-digit number. (Unit 5)
- determine whether a given whole number in the range 1–100 is prime or composite. (Unit 5)
- multiply a whole number of up to four digits by a 1-digit whole number, and multiply two 2-digit numbers. (Unit 6)
- find whole-number quotients and remainders with up to four-digit dividends and one-digit divisors. (Unit 7)

Fractions

Students

- use fraction models to explain why two fractions are equivalent, and generate equivalent fractions. (Unit 8)
- compare two fractions using benchmark fractions or by generating equivalent fractions. (Unit 8)
- decompose a fraction or mixed number into a sum of fractions with the same denominator. (Units 9, 10)
- add and subtract fractions, and mixed numbers with like denominators. (Units 9, 10)

Fractions (Continued)

- multiply a fraction or a mixed number. (Unit 11)
- represent fractions with denominators of 10 or 100 using decimal notation and compare two decimals to hundredths. (Unit 12)
- add fractions with denominators 10 and 100 by using equivalent fractions. (Unit 12)

Measurement

Students

- convert larger units of measurement to smaller equivalent units. (Unit 13)
- determine and apply the formulas for the area and perimeter of a rectangle. (Unit 13)

Data

Students

- display and interpret measurement data in line plots to solve problems. (Unit 13)

Analyze and Classify Geometric Shapes

Students

- identify and draw points, lines, line segments, and rays. (Unit 14)
- classify angles as right, acute, or obtuse, and measure and draw angles. (Unit 14)
- draw perpendicular and parallel lines and identify them in 2-dimensional figures. (Unit 14)
- recognize that when an angle is decomposed into parts, the angle measure of the whole is the sum of the angle measure of the parts. (Unit 14)
- classify 2-dimensional figures by the presence or absence of parallel and perpendiculars lines, or the presence or absence of angles of a specified size. (Unit 14)
- recognize a line of symmetry for a 2-dimensional figure. (Unit 14)
- explain how to find lines of symmetry on 2-dimensional figures. (Unit 14)

Grade 5

Habits of Mind and Classroom Norms for Productive Math Learning

Students

- make sense of problems and quantities and represent them different ways. (Unit 1)
- represent a real-world situation using mathematics. (Unit 1)
- construct an argument to explain their thinking with clear and appropriate terms. (Unit 1)
- use patterns to develop efficient strategies to solve problems. (Unit 1)
- tell their math biographies.
- recognize the behaviors and attitudes that support a productive learning environment. (Unit 1)

Operations with Fractions

Students

- add, subtract, and multiply fractions, including mixed numbers, with unlike denominators. (Units 9, 10)
- find the area of a rectangle with fractional side lengths. (Unit 10)
- describe multiplication as scaling. (Unit 10)
- divide unit fractions by whole numbers and whole numbers by unit fractions. (Unit 11)

Operations with Whole Numbers and Decimals

Students

- describe the relationship among place value positions. (Unit 3)
- use an algorithm to multiply whole numbers. (Unit 5)
- divide multi-digit dividends by 2-digit divisors. (Unit 7)
- add, subtract, multiply, or divide decimals. (Units 4, 6, 8)
- solve word problems involving operations with whole numbers or decimals. (Units 4, 5, 6, 7, 8)

Measurement

Students

- describe volume is an attribute of solid figures. (Unit 2)
- measure volume by counting unit cubes. (Unit 2)
- calculate the volume of rectangular prisms using formulas. (Unit 2)
- find the volume of composite solid figures. (Unit 2)
- convert measurement units within a given measurement system. (Unit 12)

Data

Students

- interpret data on a line plot. (Unit 12)

Geometry

Students

- identify and describe features of a coordinate plane. (Unit 13)
- graph points on the coordinate plane to solve problems. (Unit 13)
- classify 2-dimensional figures into categories based on their properties. (Unit 13)

Algebraic Thinking

Students

- write numerical expressions to represent calculations that are described using written statements. (Unit 14)
- interpret numerical expressions without evaluating them. (Unit 14)
- use the order of operations to evaluate numerical expressions. (Unit 14)
- generate two numerical patterns using two given rules. (Unit 14)
- identify apparent relationships between corresponding terms in the generated number patterns. (Unit 14)

Scope and Sequence

This scope and sequence shows the progression of concepts and skills for each domain of mathematical content: Counting and Cardinality (Kindergarten only), Operations and Algebraic Thinking, Number and Operations in Base Ten, Number and Operations—Fraction (Grades 3–5 only), Measurement, Data, and Geometry.

The numbers in the cells indicate the unit(s) that address the concepts and skills listed.

	Grade K Units	Grade 1 Units	Grade 2 Units	Grade 3 Units	Grade 4 Units	Grade 5 Units
Counting and Cardinality						
Know number names and the count sequence.						
Count to 10 by ones	12					
Count to 10 by tens.	12					
Count forward from a given number	12					
Write numbers from 0 to 20.	3, 9					
Represent up to 20 objects with a written numeral.	9, 10					
Count to tell the number of objects.						
Understand the relationship between numbers and quantities.	2, 3					
Connect counting to cardinality	2, 3					
Count objects, saying the number names in the standard order	2, 3					
Pair each object counted with one and only one number name and vice versa	2, 3					
Understand that each successive number name represents one more.	2, 3					
Understand that the last number said tells the number of objects in a group.	2, 3					
Understand that the number of objects in a given group is the same regardless of their arrangement.	2, 3					

	Grade K Units	Grade 1 Units	Grade 2 Units	Grade 3 Units	Grade 4 Units	Grade 5 Units
Count to tell the number of objects.						
Count to know how many objects in a group of up to 10 objects in a scattered formation.	2, 3					
Count to know how many objects in a group of up to 20 objects in a line, rectangular array, or circle.	2, 3, 9, 10					
Given a number up to 20, count out that many objects.	9, 10, 12					
Compare numbers.						
Compare the number of objects in two groups using matching or counting.	2, 3					
Compare two numbers between 1 and 10.	2, 3					

	Grade K Units	Grade 1 Units	Grade 2 Units	Grade 3 Units	Grade 4 Units	Grade 5 Units
Operations and Algebraic Thinking						
Understand addition.						
Represent addition using a range of models.	5, 6, 7					
Represent subtraction using a range of models.	5, 6, 7					
Add within 10 using objects and drawings	5, 6, 7					
Subtract within 10 using objects and drawings	5, 6, 7					
Solve addition problems within 10.	5, 6, 7					
Solve subtraction problems within 10.	5, 6, 7					
Decompose numbers up to 10 in multiple ways.	5, 6, 7					
Make a 10 using objects and drawings	5, 6, 7					
Fluently add within 5	5, 6, 7					
Fluently subtraction within 5	5, 6, 7					
Represent and solve problems involving addition and subtraction.						
Solve addition problems within 20.		4, 7, 8, 10				
Solve subtraction problems within 20.		4, 7, 8, 10				
Solve addition problems within 20 with 3 addends.		4, 7, 8, 10				
Apply properties of operations and the relate addition and subtraction.						
Use properties of operations to add.		4, 5				
Understand subtraction as an unknown addend problem.		4, 5				
Fluently add and subtract.						
Relate counting to addition		4, 5				
Relate counting to subtraction		4, 5				
Add within 20 using different strategies		4, 5				
Subtract within 20 using different strategies		4, 5				
Fluently add within 10.		4, 5				
Fluently subtract within 10.		4, 5				

	Grade K Units	Grade 1 Units	Grade 2 Units	Grade 3 Units	Grade 4 Units	Grade 5 Units
Fluently add within 20.			5, 6			
Fluently subtract within 20.			5, 6			
Work with addition and subtraction equations.						
Understand the meaning of the equal sign.		4, 5				
Determine whether an addition equation is true.		4, 5				
Determine whether a subtraction equation is true.		4, 5				
Determine the unknown in an addition equation.		4, 5				
Determine the unknown in a subtraction equation.		4, 5				
Represent and solve problems involving addition and subtraction.						
Add within 100 to solve one-step problems.			3, 6, 10			
Subtract within 100 to solve one-step problems.			3, 6, 10			
Add within 100 to solve two-step problems.			3, 6, 10			
Subtract within 100 to solve two-step problems.			3, 6, 10			
Work with equal groups of objects to gain foundations for multiplication.						
Determine whether a group of objects has an even or odd number of objects.			2			
Use addition to find the total number of objects arranged in a rectangular array.			2			
Represent and solve problems involving multiplication and division.						
Understand multiplication as the product of the number of equal groups of objects.				3, 4, 5, 11		
Understand division as the partitioning of a group of objects into smaller equal groups.				3, 4, 5, 11		
Multiply within 100 to solve problems.				3, 4, 5, 11		
Divide within 100 to solve problems.				3, 4, 5, 11		

	Grade K Units	Grade 1 Units	Grade 2 Units	Grade 3 Units	Grade 4 Units	Grade 5 Units
Determine the unknown in a multiplication equation.				3, 4, 5, 11		
Determine the unknown in a division equation.				3, 4, 5, 11		
Understand properties of multiplication and the relationship between multiplication and division.						
Use properties of operations to multiply.				3, 5, 9, 10		
Understand division as an unknown-factor problem.				3, 5, 9, 10		
Multiply and divide within 100.						
Fluently multiply within 100.				4, 5, 9		
Fluently divide within 100.				4, 5, 9		
Solve problems involving the four operations, and identify and explain patterns in arithmetic						
Solve two-step problems using four operations.				2, 4, 10		
Represent the unknown in an equation with a letter.				2, 4, 10	3, 4, 6, 7, 13	
Assess the reasonableness of answer using estimation strategies.				2, 4, 10	3, 4, 6, 7, 13	
Identify arithmetic patterns				2, 4, 10		
Explain arithmetic patterns using properties of operations.				2, 4, 10		
Use the four operations with whole numbers to solve problems.						
Interpret multiplication as a comparison.					3, 4, 6, 7, 13	
Solve problems involving multiplicative comparison.					3, 4, 6, 7, 13	
Distinguish multiplicative comparison from additive comparison.					3, 4, 6, 7, 13	
Solve multistep problems with whole numbers using four operations.					3, 4, 6, 7, 13	
Gain familiarity with factors and multiples						
Find all factor pairs for a whole number up to 100.					5	
Understand that a whole number is a multiple of each of its factors.					5	
Determine whether a given number is prime or composite.					5	

	Grade K Units	Grade 1 Units	Grade 2 Units	Grade 3 Units	Grade 4 Units	Grade 5 Units
Generate and analyze patterns.						
Generate a number pattern that follows a give rule.					5, 8	
Generate a shape pattern that follows a give rule.					5, 8	
Identify apparent features of a pattern that are not explicit in the rule.					5, 8	
Write and interpret numerical expressions.						
Use parentheses, brackets, or braces in numerical expressions.						14
Evaluate expressions with parentheses, brackets or braces.						14
Analyze patterns and relationships.						
General two numerical patterns using two given rules.						14
Identify apparent relationships between corresponding terms.						14
Form ordered pairs from the two patterns.						14
Graph ordered pairs on a coordinate plan.						14

	Grade K Units	Grade 1 Units	Grade 2 Units	Grade 3 Units	Grade 4 Units	Grade 5 Units
Number and Operations in Base Ten						
Work with numbers 11–19 to gain foundations for place value.						
Compose numbers from 11 to 19.	10					
Decompose numbers from 11 to 19.	10					
Understand that teen numbers are composed of ten ones and some more ones.	10					
Extend the counting sequence.						
Count to 120 starting at any number less than 120.		2				
Read and write numerals.		2				
Represent a number of objects with a written numeral.		2				
Understand place value.						
Count within 1000.			2, 4			
Skip count by 5, 10, and 100.			2, 4			
Understand that the 2 digits in a 2-dgit number represent some tens and some ones.		3				
Understand that the 3 digits in a 3-dgit number represent some hundreds, tens and ones.			2, 4			
Understand the structure of base-ten place value system.					2, 6, 7	3, 8
Use whole number exponents to denote powers of 10.						3, 8
Explain patterns in the number of zeros of a product when multiplying a number by a power of 10.						3, 8
Explain patterns in the placement of the decimal points when a decimal is multiplied or divided by a power of 10.						3, 8
Compare Numbers						
Compare two 2-digit numbers based on place value.		3				
Compare two 3-digit numbers based on place value.			2, 4			

	Grade K Units	Grade 1 Units	Grade 2 Units	Grade 3 Units	Grade 4 Units	Grade 5 Units
Compare two multi-digit whole numbers.					2, 6, 7	
Compare two decimals to thousandths.						3, 8
Round Numbers						
Round whole numbers to nearest 10 or 100.				2, 10		
Round multi-digit whole numbers to any place.					2, 6, 7	
Round decimals to any place.						3, 8
Use place value understanding and properties of operations to perform multi-digit arithmetic.						
Mentally find 10 more or 10 less than a given number.		9, 11				
Mentally add or subtract 10 or 100 to a given number.			5, 6, 9, 10			
Add within 100 using a range of strategies.		9, 11				
Add within 1000 using a range of strategies.			5, 6, 9, 10			
Fluently add within 1000.				2, 10		
Fluently add multi-digit whole numbers using the standard algorithm.					3, 6, 7	
Subtract multiples of 10 from numbers up to 100.		9, 11				
Subtract within 1000 using a range of strategies.			5, 6, 9, 10			
Fluently subtract within 1000.				2, 10		
Fluently subtract multi-digit whole numbers using the standard algorithm.					3, 6, 7	
Explain addition and subtraction strategies using place value and properties of operations.			5, 6, 9, 10			
Multiply 1-digit number by multiples of 10				2, 10		
Multiply a whole number of up to 4 digits by a 1-digit whole number.					3, 6, 7	
Multiply two 2-digit numbers.					3, 6, 7	
Fluently multiply multi-digit whole numbers using the standard algorithm.						4, 5, 6, 7, 8

	Grade K Units	Grade 1 Units	Grade 2 Units	Grade 3 Units	Grade 4 Units	Grade 5 Units
Find whole number quotients and remainders with up to 4-digit dividends and one-digit divisors.					3, 6, 7	
Find whole number quotients and remainders with up to 4-digit dividends and 2-digit divisors						4, 5, 6, 7, 8
Understand the place value system.						
Read and write multi-digit whole numbers.					2, 6, 7	
Read and write decimals to thousandths.						3, 8
Perform operations with decimals to hundredths.						
Fluently multiply multi-digit whole numbers using the standard algorithm.						4, 5, 6, 7, 8
Add, subtract, multiply, and divide decimals to hundredths						4, 5, 6, 7, 8

	Grade K Units	Grade 1 Units	Grade 2 Units	Grade 3 Units	Grade 4 Units	Grade 5 Units
Number and Operations—Fractions						
Develop understanding of fractions as numbers.						
Understand what a fraction is.				7, 8		
Represent fractions on the number line.				7, 8		
Relate whole numbers and fractions.				7, 8		
Fraction equivalence						
Explain equivalence of fractions.				7, 8		
Relate fraction equivalence to size.				7, 8		
Relate fraction equivalence to the number line.				7, 8		
Generate equivalent fractions.				7, 8	8	
Explain fraction equivalence.					8	
Express fractions with denominator 10 as equivalent fractions with denominator 100.					12	
Compare fractions						
Compare fractions with the same denominator by reasoning about their size.				7, 8		
Compare fractions with the same numerator by reasoning about their size.				7, 8		
Compare fractions with different numerators.					8	
Compare fraction with different denominators.					8	
Operations with Fractions						
Add fractions with like denominators.					9, 10, 11	
Add mixed numbers with like denominators.					9, 10, 11	
Subtract fractions with like denominators.					9, 10, 11	
Subtract mixed numbers with like denominators.					9, 10, 11	
Solve problems involving addition and subtraction of fractions.					9, 10, 11	

	Grade K Units	Grade 1 Units	Grade 2 Units	Grade 3 Units	Grade 4 Units	Grade 5 Units
Add fractions with unlike denominators.						9
Subtract fractions with unlike denominators.						9
Solve problems involving addition of fractions with unlike denominators.						9
Solve problem involving subtraction of fractions with unlike denominators.						9
Multiply a fraction by a whole number.					9, 10, 11	
Multiply fractions.						10, 11
Interpret multiplication as scaling.						10, 11
Solve problems involving multiplication of fractions.					9, 10, 11	10, 11
Solve problems involving division of whole numbers with quotients that are fractions.						10, 11
Divide fractions by whole numbers and whole numbers by fractions.						10, 11
Understand decimal notation for fractions and compare decimal fractions.						
Add two fractions with denominators 10 and 100.					12	
Write decimal fractions using decimal notation.					12	
Compare two decimals to hundredths by reasoning about their size.					12	

Measurement

	Grade K Units	Grade 1 Units	Grade 2 Units	Grade 3 Units	Grade 4 Units	Grade 5 Units
Describe and compare measurable attributes.						
Describe measurable attributes of objects, such as length or weight.	14					
Compare two objects for the same measurable attribute.	14					
Measure and estimate lengths						
Order three objects by length.		12				
Compare the length of two objects indirectly by comparing to the length of a third object.		12				
Measure the length of an object using an appropriate tool.			7			
Measure the length of an object using two different units.			7			
Estimate lengths of objects.			7			
Compare the lengths of two objects.			7			
Relate addition and subtraction to length.						
Use addition within 100 to solve problems involving length.			7, 9			
Use subtraction within 100 to solve problems involving length.			7, 9			
Represent whole numbers as lengths on a number line.			7, 9			
Show sums and differences within 100 on a number line.			7, 9			
Solve problems involving measurement and conversion of measurements.						
Know relative sizes of measurement within one system of measurement.					13	
Express measurements in a larger unit in terms of a smaller unit.					13	
Record measurement equivalents in a two-column table.					13	
Solve problems involving distances, intervals of time, liquid volumes, masses of objects, and money.					13	
Represent measurement quantities using diagrams, such as number line diagrams.					13	

Scope and Sequence 105

	Grade K Units	Grade 1 Units	Grade 2 Units	Grade 3 Units	Grade 4 Units	Grade 5 Units
Convert like measurement within a given measurement system.						
Convert among different-sized standard measurement within a given system.						12
Solve multi-step problems involving conversions.						12
Work with time and money.						
Tell and write time in hours using analog and digital clocks.		12				
Tell and write time in half-hours using analog and digital clocks.		12				
Tell and write time to the nearest five minutes on analog and digital clocks.			8			
Tell and write time to the nearest minute.				12		
Measure time interval in minutes				12		
Solve problems involving addition of time intervals in minutes				12		
Solve problems involving subtraction of time intervals in minutes				12		
Solve problems involving bills and coins.			8			
Solve problems involving measurement and estimation of liquid volumes and masses of objects.						
Measure liquid volume and masses of objects.				12		
Estimate liquid volume and masses of objects.				12		
Solve one-step problems involving liquid volumes.				12		
Solve one-step problems involving masses.				12		
Geometric measurement: understand concepts of area.						
Understand area as an attribute of plane figures.				6		
Understand concepts of area measurement.				6		
Measure the area of a rectangle by counting tiles.				6		

	Grade K Units	Grade 1 Units	Grade 2 Units	Grade 3 Units	Grade 4 Units	Grade 5 Units
Use multiplication to determine the area of a rectangle.				6		
Use area models to represent the distributive property.				6		
Find the area of rectilinear figures by decomposing them into rectangles and adding the areas of the rectangles.				6		
Solve problems involving the area of rectilinear figures.				6		
Apply the area formula for rectangles to solve problems.					13	
Geometric measurement: understand perimeter						
Find the perimeter of polygons given side lengths.				11		
Determine an unknown side length given the perimeter and other side lengths.				11		
Show rectangles with the same perimeter and different areas.				11		
Show rectangles with the same area and different perimeters.				11		
Solve problems involving perimeters of polygons.				11		
Apply the perimeter formula for rectangles to solve problems.					13	
Geometric measurement: understand concepts of angle and measure angles.						
Understand that angles are geometric shapes.					14	
Understand concepts of angle measurement.					14	
Measure angles in whole-number degrees using a protractor.					14	
Sketch angles of specified measure.					14	
Understand angle measure as additive.					14	
Solve addition problems to find unknown angles.					14	
Solve subtraction problems to find unknown angles.					14	

Scope and Sequence

	Grade K Units	Grade 1 Units	Grade 2 Units	Grade 3 Units	Grade 4 Units	Grade 5 Units
Geometric measurement: understand concepts of volume.						
Understand volume as an attribute of 3-dimensional figures.						2
Understand concepts of volume measurement						2
Measure volume by counting cubes.						2
Find the volume of a right rectangular prism by multiplying the edge lengths.						2
Represent three-fold whole number products as volumes to show the associative property.						2
Use the volume formula to determine volume.						2
Understand that volume is additive.						2
Find volumes of composite 3-dimensional figures						2
Solve problems involving volume.						2

Data

	Grade K Units	Grade 1 Units	Grade 2 Units	Grade 3 Units	Grade 4 Units	Grade 5 Units
Classify objects and count the number of objects in categories.						
Classify object into given categories	4					
Count the number of objects in each category.	4					
Represent and interpret data.						
Organize, represent, and interpret data with up to three categories.		12				
Analyze data by determining total number of data points, the number in each category.		12				
Compare the number of data points in different categories.		12				
Generate measurement data of lengths of object.			11	12		
Make a line plot to show measurement data			11			
Make a line plot with fractional intervals to display measurement data gathered.				12	13	12
Solve problems involving addition and subtraction of fractions using information presented in line plots					13	
Solve problems involving information presented in line plots with fractional values.						12
Draw a picture graph to represent a data set.			11			
Draw a bar graph to represent a data set.			11			
Solve problems about the data presented in a bar graph.			11			
Draw a scaled picture graph to represent a data set.				12		
Draw a scaled bar graph to represent a data set.				12		
Solve one- and two-step problems using information presented in scaled bar graphs.				12		

	Grade K Units	Grade 1 Units	Grade 2 Units	Grade 3 Units	Grade 4 Units	Grade 5 Units
Geometry						
Identify and describe shapes.						
Describe shapes in the environment.	8, 11					
Describe position of objects relative to other objects.	8, 11					
Recognize and name shapes with different sizes and orientations.	8, 11					
Understand that 2-dimensional figures are flat.	8, 11					
Understand that 3-dimensional figures are solid.	8, 11					
Analyze, compare, create, and compose shapes.						
Analyze and compare 2-dimensional figures.	13					
Analyze and compare 3-dimensional figures.	13					
Build and draw shapes that can be found in the world.	13					
Compose simple shapes to form other shapes.	13					
Reason with shapes and their attributes.						
Distinguish between defining and non-defining attributes.		6, 13				
Build or draw shapes with given defining attributes.		6, 13				
Compose 2-dimensional and 3-dimensional figures.		6, 13				
Compose new shapes from composite shapes.		6, 13				
Recognize and draw 2-dimensional and 3-dimensional figures with specified attributes.			12			
Identify triangles, quadrilaterals, pentagons, hexagon, and cubes.			12			
Understand that shapes in different categories may share attributes.				7, 13		
Understand that shared attributes of shapes can define a larger category.				7, 13		

	Grade K Units	Grade 1 Units	Grade 2 Units	Grade 3 Units	Grade 4 Units	Grade 5 Units
Recognize rhombuses, rectangles, and squares as examples of quadrilaterals.				7, 13		
Classify two-dimensional figures in a hierarchy based on properties.						13
Use a Venn diagram to organize 2-dimensional figures based on attributes.						13
Partition shapes into equal parts.						
Partition circles and rectangles into two, three, or four equal parts.		6, 13	12			
Understand that decomposing shapes into more equal parts creates smaller parts.		6, 13				
Partition a rectangle into rows and columns of the same-size squares.			12			
Recognize that equal parts of identical wholes do not always have the same shape.			12			
Partition shapes into parts with equal areas.				7, 13		
Express the area of each equal part of a shape as a fraction of the whole.				7, 13		
Draw and identify lines and angles and classify shapes by properties of their lines and angles.						
Draw and identify points, lines, line segments, rays, and angles.					14	
Draw and identify parallel and perpendicular lines					14	
Draw and identify right, acute, and obtuse angles.					14	
Use angle measure to classify figures					14	
Identify figures with line symmetry.					14	
Draw lines of symmetry.					14	
Understand the coordinate system.						
Understand a coordinate system						13
Graph points on the first quadrant of the coordinate plane.						13
Interpret coordinate values of points in the first quadrant of the coordinate plane.						13

Pacing

Kindergarten

UNIT 1 — 10 Days
Math is...
- 1-1 Math Is Mine
- 1-2 Math Is Exploring and Thinking
- 1-3 Math Is In My World
- 1-4 Math Is Explaining and Sharing
- 1-5 Math Is Finding Patterns
- 1-6 Math Is Ours

UNIT 2 — 15 Days
Numbers to 5
- 2-1 Count 1, 2, and 3
- 2-2 Represent 1, 2, and 3
- 2-3 Count 4 and 5
- 2-4 Represent 4 and 5
- 2-5 Represent 0
- 2-6 Numbers to 5
- 2-7 Equal Groups to 5
- 2-8 Greater Than and Less Than
- 2-9 Compare Numbers to 5

UNIT 3 — 18 Days
Numbers to 10
- 3-1 Count 6 and 7
- 3-2 Represent 6 and 7
- 3-3 Count 8 and 9
- 3-4 Represent 8 and 9
- 3-5 Count 10
- 3-6 Represent 10
- 3-7 Numbers to 10
- 3-8 Compare Objects in Groups
- 3-9 Compare Numbers
- 3-10 Write Numbers to 3
- 3-11 Write Numbers to 6
- 3-12 Write Numbers to 10

UNIT 4 — 8 Days
Sort, Classify, and Count Objects
- 4-1 Alike and Different
- 4-2 Sort Objects into Groups
- 4-3 Count Objects in Groups
- 4-4 Describe Groups of Objects

UNIT 5 — 9 Days
2-Dimensional Shapes
- 5-1 Triangles
- 5-2 Squares and Rectangles
- 5-3 Hexagons
- 5-4 Circles
- 5-5 Position of 2-Dimensional Shapes

UNIT 6 — 9 Days
Understand Addition
- 6-1 Represent and Solve Add To Problems
- 6-2 Represent and Solve More Add to Problems
- 6-3 Represent and Solve Put Together Problems
- 6-4 Represent and Solve Addition Problems
- 6-5 Represent and Solve More Addition Problems

UNIT 7 — 9 Days
Understand Subtraction
- 7-1 Represent Take Apart Problems
- 7-2 Represent and Solve Take From Problems
- 7-3 Represent and Solve More Take From Problems
- 7-4 Represent and Solve Subtraction Problems
- 7-5 Represent and Solve Addition and Subtraction Problems

UNIT 8:   14 Days
Addition and Subtraction Strategies
- 8-1   Add within 5
- 8-2   Subtract within 5
- 8-3   Ways to Make 6 and 7
- 8-4   Ways to Decompose 6 and 7
- 8-5   Ways to Make 8 and 9
- 8-6   Ways to Decompose 8 and 9
- 8-7   Ways to Make 10
- 8-8   Ways to Decompose 10

UNIT 9:   10 Days
Numbers 11 to 15
- 9-1   Represent 11, 12, and 13
- 9-2   Make 11, 12, and 13
- 9-3   Decompose 11, 12, and 13
- 9-4   Represent 14 and 15
- 9-5   Make 14 and 15
- 9-6   Decompose 14 and 15

UNIT 10:   10 Days
Numbers 16 to 19
- 10-1   Represent 16 and 17
- 10-2   Make 16 and 17
- 10-3   Decompose 16 and 17
- 10-4   Represent 18 and 19
- 10-5   Make 18 and 19
- 10-6   Decompose 18 and 19

UNIT 11:   10 Days
3-Dimensional Shapes
- 11-1   2-Dimensional and 3-Dimensional Shapes
- 11-2   Cubes
- 11-3   Spheres
- 11-4   Cylinders
- 11-5   Cones
- 11-6   Describe 3-Dimensional Shapes

UNIT 12:   9 Days
Count to 100
- 12-1   Count by 1s to 50
- 12-2   Count by 1s to 100
- 12-3   Count by 10s to 100
- 12-4   Count From Any Number to 100
- 12-5   Count to Find Out How Many

UNIT 13:   10 Days
Analyze, Compare, and Compose Shapes
- 13-1   Compare and Contrast 2-Dimensional Shapes
- 13-2   Build and Draw 2-Dimensional Shapes
- 13-3   Compose 2-Dimensional Shapes
- 13-4   Compare and Contrast 3-Dimensional Shapes
- 13-5   Build 3-Dimensional Shapes
- 13-6   Describe 3-Dimensional Shapes in the World

UNIT 14:   9 Days
Compare Measurable Attributes
- 14-1   Describe Attributes of Objects
- 14-2   Compare Lengths
- 14-3   Compare Heights
- 14-4   Compare Weights
- 14-5   Compare Capacity

Grade 1

UNIT 1:   **10 Days**
Math Is...
- 1-1 Math Is Mine
- 1-2 Math Is Exploring and Thinking
- 1-3 Math Is In My World
- 1-4 Math Is Explaining and Sharing
- 1-5 Math Is Finding Patterns
- 1-6 Math Is Ours

UNIT 2:   **9 Days**
Number Patterns
- 2-1 Counting Patterns to 100
- 2-2 Patterns on a Number Chart to 120
- 2-3 Patterns on a Number Line
- 2-4 Patterns When Reading and Writing Numbers
- 2-5 Patterns When Representing Objects in a Group

UNIT 3:   **14 Days**
Place Value
- 3-1 Numbers 11 to 19
- 3-2 Understand Tens
- 3-3 Represent Tens and Ones
- 3-4 Represent 2-Digit Numbers
- 3-5 Represent 2-Digit Numbers in Different Ways
- 3-6 Compare Numbers
- 3-7 Compare Numbers on a Number Line
- 3-8 Use Symbols to Compare Numbers

UNIT 4:   **17 Days**
Addition within 20: Facts and Strategies
- 4-1 Relate Counting to Addition
- 4-2 Count On to Add
- 4-3 Doubles
- 4-4 Near Doubles
- 4-5 Make a 10 to Add
- 4-6 Choose Strategies to Add
- 4-7 Use Properties to Add
- 4-8 Add Three Numbers
- 4-9 Find an Unknown Number in an Addition Equation
- 4-10 Understand the Equal Sign
- 4-11 True Addition Equations

UNIT 5:   **15 Days**
Subtraction within 20: Facts and Strategies
- 5-1 Relate Counting to Subtraction
- 5-2 Count Back to Subtract
- 5-3 Count On to Subtract
- 5-4 Make a 10 to Subtract
- 5-5 Use Near Doubles to Subtract
- 5-6 Use Addition to Subtract
- 5-7 Use Fact Families to Subtract
- 5-8 Find an Unknown Number in a Subtraction Equation
- 5-9 True Subtraction Equations

UNIT 6:   **10 Days**
Shapes and Solids
- 6-1 Understand Defining Attributes of Shapes
- 6-2 Understand Non-Defining Attributes
- 6-3 Compose Shapes
- 6-4 Build New Shapes
- 6-5 Understand Attributes of Solids
- 6-6 Build New Solids

UNIT 7:   **10 Days**
Meanings of Addition
- 7-1 Represent and Solve Add To Problems
- 7-2 Represent and Solve More Add To Problems
- 7-3 Represent and Solve Put Together Problems
- 7-4 Represent and Solve More Put Together Problems
- 7-5 Represent and Solve Addition Problems with Three Addends
- 7-6 Solve Addition Problems

UNIT 8: Meanings of Subtraction — 12 Days

- **8-1** Represent and Solve Take From Problems
- **8-2** Represent and Solve More Take From Problems
- **8-3** Represent and Solve Take Apart Problems
- **8-4** Represent and Solve More Take Apart Problems
- **8-5** Solve Problems Involving Subtraction
- **8-6** Solve More Problems Involving Subtraction
- **8-7** Solve Problems Involving Addition and Subtraction

UNIT 9: Addition within 100 — 14 Days

- **9-1** Use Mental Math to Find 10 More
- **9-2** Represent Adding Tens
- **9-3** Represent Adding Tens and Ones
- **9-4** Decompose Addends to Add
- **9-5** Use an Open Number Line to Add within 100
- **9-6** Decompose to Add on an Open Number Line
- **9-7** Regroup to Add
- **9-8** Add 2-Digit Numbers

UNIT 10: Compare Using Addition and Subtraction — 8 Days

- **10-1** Represent and Solve Compare Problems
- **10-2** Represent and Solve Compare Problems Using Addition
- **10-3** Represent and Solve More Compare Problems
- **10-4** Solve Compare Problems Using Addition and Subtraction

UNIT 11: Subtraction within 100 — 10 Days

- **11-1** Use Mental Math to Find 10 Less
- **11-2** Represent Subtracting Tens
- **11-3** Subtract Tens
- **11-4** Use Addition to Subtract Tens
- **11-5** Explain Subtraction Strategies

UNIT 12: Measurement and Data — 16 Days

- **12-1** Compare and Order Lengths
- **12-2** More Ways to Compare Lengths
- **12-3** Strategies to Measure Lengths
- **12-4** More Strategies to Measure Lengths
- **12-5** Tell Time to the Hour
- **12-6** Tell Time to the Half Hour
- **12-7** Organize Data
- **12-8** Represent Data
- **12-9** Interpret Data
- **12-10** Solve Problems Involving Data

UNIT 13: Equal Shares — 10 Days

- **13-1** Understand Equal Shares
- **13-2** Partition Shapes into Halves
- **13-3** Partition Shapes Into Fourths
- **13-4** Describe the Whole
- **13-5** Describe Halves and Fourths of Shapes

Grade 2

UNIT 1:   **10 Days**
Math Is..
1-1   Math Is Mine
1-2   Math Is Exploring and Thinking
1-3   Math Is In My World
1-4   Math Is Explaining and Sharing
1-5   Math Is Finding Patterns
1-6   Math Is Ours

UNIT 2:   **9 Days**
Place Value to 1,000
2-1   Understand Hundreds
2-2   Understand 3-Digit Numbers
2-3   Read and Write Numbers to 1,000
2-4   Decompose 3-Digit Numbers
2-5   Compare 3-Digit Numbers

UNIT 3:   **12 Days**
Patterns within Numbers
3-1   Counting Patterns
3-2   Patterns When Skip Counting by 5s
3-3   Patterns When Skip Counting by 10s and 100s
3-4   Understand Even and Odd Numbers
3-5   Addition Patterns
3-6   Patterns with Arrays
3-7   Use Arrays to Add

UNIT 4:   **16 Days**
Meanings of Addition and Subtraction
4-1   Represent and Solve Add To Problems
4-2   Represent and Solve Take From Problems
4-3   Solve Two-Step Add To and Take From Problems
4-4   Represent and Solve Put Together Problems
4-5   Represent and Solve Take Apart Problems
4-6   Solve Two-Step Put Together and Take Apart Problems
4-7   Represent and Solve Compare Problems
4-8   Represent and Solve More Compare Problems
4-9   Solve Two-Step Problems with Comparison
4-10   Solve Two-Step Problems Using Addition and Subtraction

UNIT 5:   **16 Days**
Strategies to Fluently Add within 100
5-1   Strategies to Add Fluently within 20
5-2   More Strategies to Add Fluently within 20
5-3   Represent Addition with 2-Digit Numbers
5-4   Use Properties to Add
5-5   Decompose Two Addends to Add
5-6   Use a Number Line to Add
5-7   Decompose One Adden to Add
5-8   Adjust Addends to Add
5-9   Add More Than Two Numbers
5-10   Solve One- and Two-Step Problems Using Addition

UNIT 6:   **16 Days**
Strategies to Fluently Subtract within 100
6-1   Strategies to Subtract Fluently within 20
6-2   More Strategies to Subtract Fluently within 20
6-3   Represent Subtraction with 2-Digit Numbers
6-4   Represent 2-Digit Subtraction with Regrouping
6-5   Use a Number Line to Subtract
6-6   Decompose Numbers to Subtract
6-7   Adjust Numbers to Subtract
6-8   Relate Addition to Subtraction
6-9   Solve One-Step Problems Using Subtraction
6-10   Solve Two-Step Problems Using Subtraction

UNIT 7: Measure and Compare Lengths — 17 Days

- 7-1 Measure Length with Inches
- 7-2 Measure Length with Feet and Yards
- 7-3 Compare Lengths Using Customary Units
- 7-4 Relate Inches, Feet, and Yards
- 7-5 Estimate Length Using Customary Units
- 7-6 Measure Length with Centimeters and Meters
- 7-7 Compare Lengths Using Metric Units
- 7-8 Relate Centimeters and Meters
- 7-9 Estimate Length Using Metric Units
- 7-10 Solve Problems Involving Length
- 7-11 Solve More Problems Involving Length

UNIT 8: Measurement: Money and Time — 10 Days

- 8-1 Understand the Values of Coins
- 8-2 Solve Money Problems Involving Coins
- 8-3 Solve Money Problems Involving Dollar Bills and Coins
- 8-4 Tell Time to the Nearest Five Minutes
- 8-5 Be Precise When Telling Time

UNIT 9: Strategies to Add 3-Digit Numbers — 12 Days

- 9-1 Use Mental Math to Add 10 or 100
- 9-2 Represent Addition with 3-Digit Numbers
- 9-3 Represent Addition with 3-Digit Numbers with Regrouping
- 9-4 Decompose Addends to Add 3-Digit Numbers
- 9-5 Decompose One Addend to Add 3-Digit Numbers
- 9-6 Adjust Addends to Add 3-Digit Numbers
- 9-7 Explain Addition Strategies

UNIT 10: Strategies to Subtract 3-Digit Numbers — 15 Days

- 10-1 Use Mental Math to Subtract 10 and 100
- 10-2 Represent Subtraction with 3-Digit Numbers
- 10-3 Decompose One 3-Digit Number to Count Back
- 10-4 Count On to Subtract 3-Digit Numbers
- 10-5 Regroup Tens
- 10-6 Regroup Tens and Hundreds
- 10-7 Adjust Numbers to Subtract 3-Digit Numbers
- 10-8 Explain Subtraction Strategies
- 10-9 Solve Problems Involving Addition and Subtraction

UNIT 11: Data Analysis — 10 Days

- 11-1 Understand Picture Graphs
- 11-2 Understand Bar Graphs
- 11-3 Solve Problems Using Bar Graphs
- 11-4 Collect Measurement Data
- 11-5 Understand Line Plots
- 11-6 Show Data on a Line Plot

UNIT 12: Geometric Shapes and Equal Shares — 10 Days

- 12-1 Recognize 2-Dimensional Shapes by Their Attributes
- 12-2 Draw 2-Dimensional Shapes from Their Attributes
- 12-3 Recognize 3-Dimensional Shapes from Their Attributes
- 12-4 Understand Equal Shares
- 12-5 Relate Equal Shares
- 12-6 Partition a Rectangle into Rows and Columns

Grade 3

UNIT 1:   10 Days
Math Is...
- 1-1   Math Is Mine
- 1-2   Math Is Exploring and Thinking
- 1-3   Math Is In My World
- 1-4   Math Is Explaining and Sharing
- 1-5   Math Is Finding Patterns
- 1-6   Math Is Ours

UNIT 2:   18 Days
Use Place Value to Fluently Add and Subtract within 1,000
- 2-1   Represent 4-Digit Numbers
- 2-2   Round Multi-Digit Numbers
- 2-3   Estimate Sums and Differences
- 2-4   Use Addition Properties to Add
- 2-5   Addition Patterns
- 2-6   Use Partial Sums to Add
- 2-7   Decompose to Subtract
- 2-8   Adjust Numbers to Add or Subtract
- 2-9   Use Addition to Subtract
- 2-10   Fluently Add within 1,000
- 2-11   Fluently Subtract within 1,000

UNIT 3:   12 Days
Multiplication and Division
- 3-1   Understand Equal Groups
- 3-2   Use Arrays to Multiply
- 3-3   Understand the Commutative Property
- 3-4   Understand Equal Sharing
- 3-5   Understand Equal Grouping
- 3-6   Relate Multiplication and Division
- 3-7   Find the Unknown

UNIT 4:   10 Days
Use Patterns to Multiply by 0, 1, 2, 5, and 10
- 4-1   Use Patterns to Multiply by 2
- 4-2   Use Patterns to Multiply by 5
- 4-3   Use Patterns to Multiply by 10
- 4-4   Patterns to Multiply by 1 and 0
- 4-5   Multiply Fluently by 0, 1, 2, 5, and 10
- 4-6   Solve Problems Involving Equal Groups

UNIT 5:   12 Days
Use Properties to Multiply by 3, 4, 6, 7, 8, and 9
- 5-1   Understand the Distributive Property
- 5-2   Use Properties to Multiply by 3
- 5-3   Use Properties to Multiply by 4
- 5-4   Use Properties to Multiply by 6
- 5-5   Use Properties to Multiply by 8
- 5-6   Use Properties to Multiply by 7 and 9
- 5-7   Solve Problems Involving Arrays

UNIT 6:   11 Days
Connect Area and Multiplication
- 6-1   Understand Area
- 6-2   Count Unit Squares to Determine Area
- 6-3   Use Multiplication to Determine Area
- 6-4   Determine the Area of a Composite Figure
- 6-5   Use the Distributive Property to Determine Area
- 6-6   Solve Area Problems

UNIT 7:   10 Days
Fractions
- 7-1   Partition Shapes into Equal Parts
- 7-2   Understand Fractions
- 7-3   Represent Fractions on a Number Line
- 7-4   Represent One Whole as a Fraction
- 7-5   Represent Whole Numbers as Fractions
- 7-6   Represent a Fraction Greater Than One on a Number Line

UNIT 8:   12 Days
Fraction Equivalence and Comparison

8-1	Understand Equivalent Fractions
8-2	Represent Equivalent Fractions
8-3	Represent Equivalent Fractions on a Number Line
8-4	Understand Fractions of Different Wholes
8-5	Compare Fractions with the Same Denominator
8-6	Compare Fractions with the Same Numerator
8-7	Compare Fractions

UNIT 9:   15 Days
Use Multiplication to Divide

9-1	Use Multiplication to Solve Division Equations
9-2	Divide by 2
9-3	Divide by 5 and 10
9-4	Understand Division with 1 and 0
9-5	Divide by 3 and 6
9-6	Divide by 4 and 8
9-7	Divide by 9
9-8	Divide by 7
9-9	Multiply and Divide Fluently within 100

UNIT 10:   10 Days
Use Properties and Strategies to Multiply and Divide

10-1	Patterns with Multiples of 10
10-2	More Multiplication Patterns
10-3	Understand the Associative Property
10-4	Two-Step Problems Involving Multiplication and Division
10-5	Solve Two-Step Problems
10-6	Explain the Reasonableness of a Solution

UNIT 11:   9 Days
Perimeter

11-1	Understand Perimeter
11-2	Determine Perimeter of Figures
11-3	Determine an Unknown Side Length
11-4	Solve Problems Involving Area and Perimeter
11-5	Solve Problems Involving Measurement

UNIT 12:   17 Days
Measurement and Data

12-1	Measure Liquid Volume
12-2	Estimate and Solve Problems with Liquid Volume
12-3	Measure Mass
12-4	Estimate and Solve Problems with Mass
12-5	Tell Time to the Nearest Minute
12-6	Solve Problems Involving Time
12-7	Understand Scaled Picture Graphs
12-8	Understand Scaled Bar Graphs
12-9	Solve Problems Involving Scaled Graphs
12-10	Measure to Halves or Fourths of an Inch
12-11	Show Measurement Data on a Line Plot

UNIT 13:   8 Days
Describe and Analyze 2-Dimensional Shapes

13-1	Describe and Classify Polygons
13-2	Describe Quadrilaterals
13-3	Classify Quadrilaterals
13-4	Draw Quadrilaterals with Specific Attributes

Grade 4

UNIT 1:		10 Days
Math Is...		
1-1	Math Is Mine	
1-2	Math Is Exploring and Thinking	
1-3	Math Is In My World	
1-4	Math Is Explaining and Sharing	
1-5	Math Is Finding Patterns	
1-6	Math Is Ours	

UNIT 2:		8 Days
Generalize Place-Value Structure		
2-1	Understand the Structure of Multi-Digit Numbers	
2-2	Read and Write Numbers to One Million	
2-3	Compare Multi-Digit Numbers	
2-4	Round Multi-Digit Number	

UNIT 3:		15 Days
Addition and Subtraction Strategies and Algorithms		
3-1	Estimate Sums or Differences	
3-2	Strategies to Add Multi-Digit Numbers	
3-3	Understand an Addition Algorithm	
3-4	Understand an Addition Algorithm Involving Regrouping	
3-5	Strategies to Subtract Multi-Digit Numbers	
3-6	Understand a Subtraction Algorithm	
3-7	Understand a Subtraction Algorithm Involving Regrouping	
3-8	Represent and Solve Multi-Step Problems	
3-9	Solve Multi-Step Problems Involving Addition and Subtraction	

UNIT 4:		8 Days
Multiplication as Comparison		
4-1	Understand Comparing with Multiplication	
4-2	Represent Comparison Problems	
4-3	Solve Comparison Problems Using Multiplication	
4-4	Solve Comparison Problems Using Division	

UNIT 5:		10 Days
Numbers and Number Patterns		
5-1	Understand Factors of a Number	
5-2	Understand Prime and Composite Numbers	
5-3	Understand Multiples	
5-4	Number or Shape Patterns	
5-5	Generate a Pattern	
5-6	Analyze Features of a Pattern	

UNIT 6:		14 Days
Multiplication Strategies with Multi-Digit Numbers		
6-1	Multiply by Multiples of 10, 100, or 1,000	
6-2	Estimate Products	
6-3	Use the Distributive Property to Multiply	
6-4	Multiply 2-Digit by 1-Digit Factors	
6-5	Multiply Multi-Digit by 1-Digit Factors	
6-6	Multiple Two Multiples of 10	
6-7	Multiply Two 2-Digit Factors	
6-8	Solve Multi-Step Problems Involving Multiplication	

UNIT 7:		14 Days
Division Strategies with Multi-Digit Dividends and 1-Digit Divisors		
7-1	Divide Multiples of 10, 100, or 1,000	
7-2	Estimate Quotients	
7-3	Find Equal Shares	
7-4	Understand Partial Quotients	
7-5	Divide 4-Digit Dividends by 1-Digit Divisors	
7-6	Understand Remainders	
7-7	Make Sense of Remainders	
7-8	Solve Mutli-Step Problems Using Division	

UNIT 8:		9 Days
Fraction Equivalence		
8-1	Equivalent Fractions	
8-2	Generate Equivalent Fractions using Models	
8-3	Generate Equivalent Fractions using Number Lines	

| 8-4 | Compare Fractions using Benchmarks |
| 8-5 | Other Ways to Compare Factions |

UNIT 9: 10 Days
Addition and Subtraction Meanings and Strategies with Fractions

9-1	Understand Decomposing Fractions
9-2	Represent Adding Fractions
9-3	Add Fractions with Like Denominators
9-4	Represent Subtracting Fractions
9-5	Subtract Fractions with Like Denominators
9-6	Solve Problems Involving Fractions

UNIT 10: 10 Days
Addition and Subtraction Strategies with Mixed Numbers

10-1	Understand Decomposing Mixed Numbers
10-2	Represent Adding Mixed Numbers
10-3	Add Mixed Numbers
10-4	Represent Subtracting Mixed Numbers
10-5	Subtract Mixed Numbers
10-6	Solve Problems Involving Mixed Numbers

UNIT 11: 9 Days
Multiply Fractions by Whole Numbers

11-1	Represent Multiplication of a Unit Fraction by a Whole Number
11-2	Understand Multiplying a Fraction by a Whole Number
11-3	Multiply a Fraction by a Whole Number
11-4	Multiply a Mixed Number by a Whole Number
11-5	Solve Problems Involving Fractions and Mixed Numbers

UNIT 12: 9 Days
Decimal Fractions

12-1	Understand Tenths and Hundredths
12-2	Understand Decimal Notation
12-3	Compare Decimals
12-4	Add Decimals Using Fractions
12-5	Solve Problems Involving Money

UNIT 13: 17 Days
Units of Measurement and Data

13-1	Relate Metric Units
13-2	Relate Customary Units of Weight
13-3	Relate Customary Units of Capacity
13-4	Convert Units of Time
13-5	Solve Problems That Involve Units of Measure
13-6	Solve More Problems That Involve Units of Measure
13-7	Solve Problems Using a Perimeter Formula
13-8	Solve Problems Using an Area Formula
13-9	Solve Problems Involving Perimeter and Area
13-10	Display and Interpret Data on a Line Plot
13-11	Solve Problems Involving Data on a Line Plot

UNIT 14: 16 Days
Geometric Figures

14-1	Understand Lines, Line Segments, and Rays
14-2	Classify Angles
14-3	Draw and Measure Angles
14-4	Understand Parallel and Perpendicular Lines
14-5	Add and Subtract Angle Measures
14-6	Solve Problems Involving Unknown Angle Measures
14-7	Classify Polygons
14-8	Classify Triangles
14-9	Understand Line Symmetry
14-10	Draw Lines of Symmetry

Grade 5

UNIT 1:	10 Days
Math Is...	

- 1-1 Math Is Mine
- 1-2 Math Is Exploring and Thinking
- 1-3 Math Is In My World
- 1-4 Math Is Explaining and Sharing
- 1-5 Math Is Finding Patterns
- 1-6 Math Is Ours

UNIT 2:	9 Days
Volume	

- 2-1 Understand Volume
- 2-2 Use Unit Cubes to Determine Volume
- 2-3 Use Formulas to Determine Volume
- 2-4 Determine the Volume of Composite Figures
- 2-5 Solve Problems Involving Volume

UNIT 3:	9 Days
Place Value and Number Relationships	

- 3-1 Generalize Place Value
- 3-2 Extend Place Value to Decimals
- 3-3 Read and Write Decimals
- 3-4 Compare Decimals
- 3-5 Use Place Value to Round Decimals

UNIT 4:	14 Days
Add and Subtract Decimals	

- 4-1 Estimate Sums and Differences of Decimals
- 4-2 Represent Addition of Decimals
- 4-3 Represent Addition of Tenths and Hundredths
- 4-4 Strategies to Add Decimals
- 4-5 Represent Subtraction of Decimals
- 4-6 Represent Subtraction of Tenths and Hundredths
- 4-7 Strategies to Subtract Decimals
- 4-8 Explain Strategies to Add and Subtract Decimals

UNIT 5:	12 Days
Multiply Multi- Digit Whole Numbers	

- 5-1 Understand Powers and Exponents
- 5-2 Patterns When Multiplying a Whole Number by Powers of 10
- 5-3 Estimate Products of Multi- Digit Factors
- 5-4 Use Area Models to Multiply Multi-Digit Factors
- 5-5 Use Partial Products to Multiply Multi-Digit Factors
- 5-6 Relate Partial Products to an Algorithm
- 5-7 Multiply Multi-Digit Factors Fluently

UNIT 6:	10 Days
Multiply Decimals	

- 6-1 Patterns When Multiplying Decimals by Powers of 10
- 6-2 Estimate Products of Decimals
- 6-3 Represent Multiplication of Decimals
- 6-4 Use an Area Model to Multiply Decimals
- 6-5 Generalizations about Multiplying Decimals
- 6-6 Explain Strategies to Multiply Decimals

UNIT 7:	12 Days
Divide Whole Numbers	

- 7-1 Division Patterns with Multi-Digit Numbers
- 7-2 Estimate Quotients
- 7-3 Relate Multiplication and Division of Multi-Digit Numbers
- 7-4 Represent Division of 2-Digit Divisors
- 7-5 Use Partial Quotients
- 7-6 Divide Multi-Digit Whole Numbers
- 7-7 Solve Problems Involving Division

UNIT 8: Divide Decimals — 10 Days

8-1	Division Patterns with Decimals and Powers of 10
8-2	Estimate Quotients of Decimals
8-3	Represent Division of Decimals by a Whole Number
8-4	Divide Decimals by Whole Numbers
8-5	Divide Whole Numbers by Decimals
8-6	Divide Decimals by Decimals

UNIT 9: Add and Subtract Fractions — 15 Days

9-1	Estimating Sums and Differences of Fractions
9-2	Represent Addition of Fractions with Unlike Denominators
9-3	Add Fractions with Unlike Denominators
9-4	Represent Subtraction of Fractions with Unlike Denominators
9-5	Subtract Fractions with Unlike Denominators
9-6	Add Mixed Numbers with Unlike Denominators
9-7	Subtract Mixed Numbers with Unlike Denominators
9-8	Add and Subtract Mixed Numbers with Regrouping
9-9	Solve Problems Involving Fractions and Mixed Numbers

UNIT 10: Multiply Fractions — 15 Days

10-1	Represent Multiplication of a Fraction by a Whole Number
10-2	Multiply a Whole Number by a Fraction
10-3	Represent Multiplication of a Fraction by a Fraction
10-4	Multiply a Fraction by a Fraction
10-5	Determine the Area of Rectangles with Fractional Side Lengths
10-6	Represent Multiplication of Mixed Numbers
10-7	Multiply Mixed Numbers
10-8	Multiplication as Scaling
10-9	Solve Problems Involving Fractions

UNIT 11: Divide Fractions — 11 Days

11-1	Relate Fractions to Division
11-2	Solve Problems Involving Division
11-3	Represent Division of Whole Numbers by Unit Fractions
11-4	Divide Whole Numbers by Unit Fractions
11-5	Represent Division of Unit Fractions by Non-Zero Whole Numbers
11-6	Divide Unit Fractions by Non-Zero Whole Numbers
11-7	Solve Problems Involving Fractions

UNIT 12: Measurement and Data — 9 Days

12-1	Convert Customary Units
12-2	Convert Metric Units
12-3	Solve Multi-Step Problems Involving Measurement Units
12-4	Represent Measurement Data on a Line Plot
12-5	Solve Problems Involving Measurement Data on Line Plots

UNIT 13: Geometry — 10 Days

13-1	Understand the Coordinate Plane
13-2	Plot Ordered Pairs on the Coordinate Plane
13-3	Represent Problems on a Coordinate Plane
13-4	Classify Triangles by Properties
13-5	Properties of Quadrilaterals
13-6	Classify Quadrilaterals by Properties

UNIT 14: Algebraic Thinking — 10 Days

14-1	Write Numerical Expressions
14-2	Interpret Numerical Expressions
14-3	Evaluate Numerical Expressions
14-4	Numerical Patterns
14-5	Relate Numerical Patterns
14-6	Graphs of Numerical Patterns

Correlations

Kindergarten

♦ Major ▲ Supporting ○ Additional

	Standard Text		RM Lesson(s)
Counting and Cardinality			
♦	K.CC.A.1	Count to 100 by ones and by tens.	12-1, 12-2, 12-3
♦	K.CC.A.2	Count forward beginning from a given number within the known sequence (instead of having to begin at 1).	12-4
♦	K.CC.A.3	Write numbers from 0 to 20. Represent a number of objects with a written numeral 0-20 (with 0 representing a count of no objects).	2-5, 3-10, 3-11, 3-12, 9-1, 9-4, 10-1, 10-4
♦	K.CC.B.4	Understand the relationship between numbers and quantities; connect counting to cardinality. a. When counting objects, say the number names in the standard order, pairing each object with one and only one number name and each number name with one and only one object. b. Understand that the last number name said tells the number of objects counted. The number of objects is the same regardless of their arrangement or the order in which they were counted. c. Understand that each successive number name refers to a quantity that is one larger.	2-1, 2-2, 2-3, 2-4, 2-6, 3-1, 3-2, 3-3, 3-4, 3-5, 3-6, 3-7
♦	K.CC.B.5	Count to answer "how many?" questions about as many as 20 things arranged in a line, a rectangular array, or a circle, or as many as 10 things in a scattered configuration; given a number from 1–20, count out that many objects.	12-5
♦	K.CC.C.6	Identify whether the number of objects in one group is greater than, less than, or equal to the number of objects in another group, e.g., by using matching and counting strategies.	2-7, 2-8, 2-9, 3-8
♦	K.CC.C.7	Compare two numbers between 1 and 10 presented as written numerals.	2-9, 3-9
Operations and Algebraic Thinking			
♦	K.OA.A.1	Represent addition and subtraction with objects, fingers, mental images, drawings, sounds (e.g., claps), acting out situations, verbal explanations, expressions, or equations.	6-1, 6-3, 6-5, 7-1, 7-2, 7-3, 7-4, 7-5, 8-3, 8-5
♦	K.OA.A.2	Solve addition and subtraction word problems, and add and subtract within 10, e.g., by using objects or drawings to represent the problem.	6-2, 6-4, 7-4, 7-5
♦	K.OA.A.3	Decompose numbers less than or equal to 10 into pairs in more than one way, e.g., by using objects or drawings, and record each decomposition by a drawing or equation (e.g., $5 = 2 + 3$ and $5 = 4 + 1$).	8-4, 8-6, 8-8

Standard Text			RM Lesson(s)
◆	K.OA.A.4	For any number from 1 to 9, find the number that makes 10 when added to the given number, e.g., by using objects or drawings, and record the answer with a drawing or equation.	8-7
◆	K.OA.A.5	Fluently add and subtract within 5.	8-1, 8-2
Number and Operations in Base Ten			
◆	K.NBT.A.1	Compose and decompose numbers from 11 to 19 into ten ones and some further ones, e.g., by using objects or drawings, and record each composition or decomposition by a drawing or equation (e.g., 18 = 10 + 8); understand that these numbers are composed of ten ones and one, two, three, four, five, six, seven, eight, or nine ones.	9-2, 9-3, 9-5, 9-6, 10-2, 10-3, 10-5, 10-6
Measurement and Data			
○	K.MD.A.1	Describe measurable attributes of objects, such as length or weight. Describe several measurable attributes of a single object.	14-1
○	K.MD.A.2	Directly compare two objects with a measurable attribute in common, to see which object has "more of"/"less of" the attribute and describe the difference.	14-2, 14-3, 14-4, 14-5
▲	K.MD.B.3	Classify objects into given categories; count the numbers of objects in each category and sort the categories by count.	4-1, 4-2, 4-3, 4-4
Geometry			
○	K.G.A.1	Describe objects in the environment using names of shapes and describe the relative positions of these objects using terms such as above, below, beside, in front of, behind, and next to.	5-5, 11-6
○	K.G.A.2	Correctly name shapes regardless of their orientations or overall size.	5-1, 5-2, 5-3, 5-4, 11-2, 11-3, 11-4, 11-5
○	K.G.A.3	Identify shapes as two-dimensional (lying in a plane, "flat") or three-dimensional ("solid").	11-1, 13-6
▲	K.G.B.4	Analyze and compare two- and three-dimensional shapes, in different sizes and orientations, using informal language to describe their similarities, differences, parts (e.g., number of sides and vertices/"corners") and other attributes (e.g., having sides of equal length).	13-1, 13-4
▲	K.G.B.5	Model shapes in the world by building shapes from components (e.g., sticks and clay balls) and drawing shapes.	13-2, 13-5
▲	K.G.B.6	Compose simple shapes to form larger shapes.	13-3

Correlations

Grade 1

◆ Major   ▲ Supporting   ○ Additional

	Standard Text		RM Lesson(s)
Operations and Algebraic Thinking			
◆	1.OA.A.1	Use addition and subtraction within 20 to solve word problems involving situations of adding to, taking from, putting together, taking apart, and comparing, with unknowns in all positions, e.g., by using objects, drawings, and equations with a symbol for the unknown number to represent the problem.	7-1, 7-2, 7-3, 7-4, 7-6, 8-1, 8-2, 8-3, 8-4, 8-5, 8-6, 8-7, 10-1, 10-2, 10-3, 10-4
◆	1.OA.A.2	Solve word problems that call for addition of three whole numbers whose sum is less than or equal to 20, e.g., by using objects, drawings, and equations with a symbol for the unknown number to represent the problem.	4-8, 7-5
◆	1.OA.B.3	Apply properties of operations as strategies to add and subtract. Examples: If $8 + 3 = 11$ is known, then $3 + 8 = 11$ is also known. (Commutative property of addition.) To add $2 + 6 + 4$, the second two numbers can be added to make a ten, so $2 + 6 + 4 = 2 + 10 = 12$. (Associative property of addition.)	4-7, 4-8
◆	1.OA.B.4	Understand subtraction as an unknown-addend problem.	5-6
◆	1.OA.C.5	Relate counting to addition and subtraction (e.g., by counting on 2 to add 2).	4-1, 5-1
◆	1.OA.C.6	Add and subtract within 20, demonstrating fluency for addition and subtraction within 10. Use strategies such as counting on; making ten (e.g., $8 + 6 = 8 + 2 + 4 = 10 + 4 = 14$); decomposing a number leading to a ten (e.g., $13 - 4 = 13 - 3 - 1 = 10 - 1 = 9$); using the relationship between addition and subtraction (e.g., knowing that $8 + 4 = 12$, one knows $12 - 8 = 4$); and creating equivalent but easier or known sums (e.g., adding $6 + 7$ by creating the known equivalent $6 + 6 + 1 = 12 + 1 = 13$).	4-2, 4-3, 4-4, 4-5, 4-6, 5-2, 5-3, 5-4, 5-5, 5-7
◆	1.OA.D.7	Understand the meaning of the equal sign and determine if equations involving addition and subtraction are true or false.	4-10, 4-11, 5-9
◆	1.OA.D.8	Determine the unknown whole number in an addition or subtraction equation relating three whole numbers.	4-9, 5-8
Number and Operations in Base Ten			
◆	1.NBT.A.1	Count to 120, starting at any number less than 120. In this range, read and write numerals and represent a number of objects with a written numeral.	2-1, 2-2, 2-3, 2-4, 2-5

126   Content Guide

Standard Text		RM Lesson(s)
◆ **1.NBT.B.2**	Understand that the two digits of a two-digit number represent amounts of tens and ones. Understand the following as special cases: a. 10 can be thought of as a bundle of ten ones—called a "ten." b. The numbers from 11 to 19 are composed of a ten and one, two, three, four, five, six, seven, eight, or nine ones. c. The numbers 10, 20, 30, 40, 50, 60, 70, 80, 90 refer to one, two, three, four, five, six, seven, eight, or nine tens (and 0 ones).	3-1, 3-2, 3-3, 3-4, `3-5
◆ **1.NBT.B.3**	Compare two two-digit numbers based on meanings of the tens and ones digits, recording the results of comparisons with the symbols >, =, and <.	3-6, 3-7, 3-8
◆ **1.NBT.C.4**	Add within 100, including adding a two-digit number and a one-digit number, and adding a two-digit number and a multiple of 10, using concrete models or drawings and strategies based on place value, properties of operations, and/or the relationship between addition and subtraction; relate the strategy to a written method and explain the reasoning used. Understand that in adding two-digit numbers, one adds tens and tens, ones and ones; and sometimes it is necessary to compose a ten.	9-2, 9-3, 9-4, 9-5, 9-6, 9-7, 9-8
◆ **1.NBT.C.5**	Given a two-digit number, mentally find 10 more or 10 less than the number, without having to count; explain the reasoning used.	9-1, 11-1
◆ **1.NBT.C.6**	Subtract multiples of 10 in the range 10-90 from multiples of 10 in the range 10-90 (positive or zero differences), using concrete models or drawings and strategies based on place value, properties of operations, and/or the relationship between addition and subtraction; relate the strategy to a written method and explain the reasoning used.	11-2, 11-3, 11-4, 11-5
Measurement and Data		
◆ **1.MD.A.1**	Order three objects by length; compare the lengths of two objects indirectly by using a third object.	12-1, 12-2
◆ **1.MD.A.2**	Express the length of an object as a whole number of length units, by laying multiple copies of a shorter object (the length unit) end to end; understand that the length measurement of an object is the number of same-size length units that span it with no gaps or overlaps. Limit to contexts where the object being measured is spanned by a whole number of length units with no gaps or overlaps.	12-3, 12-4

	Standard Text		RM Lesson(s)
○	**1.MD.B.3**	Tell and write time in hours and half-hours using analog and digital clocks.	12-5, 12-6
▲	**1.MD.C.4**	Organize, represent, and interpret data with up to three categories; ask and answer questions about the total number of data points, how many in each category, and how many more or less are in one category than in another.	12-7, 12-8, 12-9, 12-10
Geometry			
○	**1.G.A.1**	Distinguish between defining attributes (e.g., triangles are closed and three-sided) versus non-defining attributes (e.g., color, orientation, overall size); build and draw shapes to possess defining attributes.	6-1, 6-2, 6-5
○	**1.G.A.2**	Compose two-dimensional shapes (rectangles, squares, trapezoids, triangles, half-circles, and quarter-circles) or three-dimensional shapes (cubes, right rectangular prisms, right circular cones, and right circular cylinders) to create a composite shape, and compose new shapes from the composite shape.	6-3, 6-4, 6-6
○	**1.G.A.3**	Partition circles and rectangles into two and four equal shares, describe the shares using the words halves, fourths, and quarters, and use the phrases half of, fourth of, and quarter of. Describe the whole as two of, or four of the shares. Understand for these examples that decomposing into more equal shares creates smaller shares.	13-1, 13-2, 13-3, 13-4, 13-5

Correlations

Grade 2

♦ Major ▲ Supporting ○ Additional

Standard Text			RM Lesson(s)
Operations and Algebraic Thinking			
♦	2.OA.A.1	Use addition and subtraction within 100 to solve one- and two-step word problems involving situations of adding to, taking from, putting together, taking apart, and comparing, with unknowns in all positions, e.g., by using drawings and equations with a symbol for the unknown number to represent the problem	4-1, 4-2, 4-3, 4-4, 4-5, 4-6, 4-7, 4-8, 4-9, 4-10, 5-10, 6-9, 6-10
♦	2.OA.B.2	Fluently add and subtract within 20 using mental strategies. By end of Grade 2, know from memory all sums of two one-digit numbers.	5-1, 5-2, 6-1, 6-2
▲	2.OA.C.3	Determine whether a group of objects (up to 20) has an odd or even number of members, e.g., by pairing objects or counting them by 2s; write an equation to express an even number as a sum of two equal addends.	3-4, 3-5
▲	2.OA.C.4	Use addition to find the total number of objects arranged in rectangular arrays with up to 5 rows and up to 5 columns; write an equation to express the total as a sum of equal addends.	3-6, 3-7
Number and Operations in Base Ten			
♦	2.NBT.A.1	Understand that the three digits of a three-digit number represent amounts of hundreds, tens, and ones; e.g., 706 equals 7 hundreds, 0 tens, and 6 ones. Understand the following as special cases: a. 100 can be thought of as a bundle of ten tens — called a "hundred." b. The numbers 100, 200, 300, 400, 500, 600, 700, 800, 900 refer to one, two, three, four, five, six, seven, eight, or nine hundreds (and 0 tens and 0 ones).	2-1, 2-2, 2-4
♦	2.NBT.A.2	Count within 1000; skip-count by 5s, 10s, and 100s.	3-1, 3-2, 3-3
♦	2.NBT.A.3	Read and write numbers to 1000 using base-ten numerals, number names, and expanded form.	2-3, 2-4
♦	2.NBT.A.4	Compare two three-digit numbers based on meanings of the hundreds, tens, and ones digits, using >, =, and < symbols to record the results of comparisons.	2-5
♦	2.NBT.B.5	Fluently add and subtract within 100 using strategies based on place value, properties of operations, and/or the relationship between addition and subtraction.	5-3, 5-4, 5-5, 5-7, 5-8, 6-3, 6-4, 6-6, 6-7, 6-8
♦	2.NBT.B.6	Add up to four two-digit numbers using strategies based on place value and properties of operations.	5-9

Standard Text			RM Lesson(s)
◆	2.NBT.B.7	Add and subtract within 1000, using concrete models or drawings and strategies based on place value, properties of operations, and/or the relationship between addition and subtraction; relate the strategy to a written method. Understand that in adding or subtracting three-digit numbers, one adds or subtracts hundreds and hundreds, tens and tens, ones and ones; and sometimes it is necessary to compose or decompose tens or hundreds.	9-2, 9-3, 9-4, 9-5, 9-6, 10-2, 10-3, 10-4, 10-5, 10-6, 10-7, 10-9
◆	2.NBT.B.8	Mentally add 10 or 100 to a given number 100–900, and mentally subtract 10 or 100 from a given number 100–900.	9-1, 10-1
◆	2.NBT.B.9	Explain why addition and subtraction strategies work, using place value and the properties of operations.	9-7, 10-8
Measurement and Data			
◆	2.MD.A.1	Measure the length of an object by selecting and using appropriate tools such as rulers, yardsticks, meter sticks, and measuring tapes.	7-1, 7-2, 7-6
◆	2.MD.A.2	Measure the length of an object twice, using length units of different lengths for the two measurements; describe how the two measurements relate to the size of the unit chosen.	7-4, 7-8
◆	2.MD.A.3	Estimate lengths using units of inches, feet, centimeters, and meters.	7-5, 7-9
◆	2.MD.A.4	Measure to determine how much longer one object is than another, expressing the length difference in terms of a standard length unit.	7-3, 7-7
◆	2.MD.B.5	Use addition and subtraction within 100 to solve word problems involving lengths that are given in the same units, e.g., by using drawings (such as drawings of rulers) and equations with a symbol for the unknown number to represent the problem.	7-10, 7-11
◆	2.MD.B.6	Represent whole numbers as lengths from 0 on a number line diagram with equally spaced points corresponding to the numbers 0, 1, 2, ..., and represent whole-number sums and differences within 100 on a number line diagram.	5-6, 6-5, 7-11
▲	2.MD.C.7	Tell and write time from analog and digital clocks to the nearest five minutes, using a.m. and p.m.	8-4, 8-5
▲	2.MD.C.8	Solve word problems involving dollar bills, quarters, dimes, nickels, and pennies, using $ and ¢ symbols appropriately.	8-1, 8-2, 8-3

Standard Text		RM Lesson(s)
▲ 2.MD.D.9	Generate measurement data by measuring lengths of several objects to the nearest whole unit, or by making repeated measurements of the same object. Show the measurements by making a line plot, where the horizontal scale is marked off in whole-number units.	11-4, 11-5, 11-6
▲ 2.MD.D.10	Draw a picture graph and a bar graph (with single-unit scale) to represent a data set with up to four categories. Solve simple put together, take-apart, and compare problems using information presented in a bar graph.	11-1, 11-2, 11-3
Geometry		
○ 2.G.A.1	Recognize and draw shapes having specified attributes, such as a given number of angles or a given number of equal faces. Identify triangles, quadrilaterals, pentagons, hexagons, and cubes.	12-1, 12-2, 12-3
○ 2.G.A.2	Partition a rectangle into rows and columns of same-size squares and count to find the total number of them.	12-6
○ 2.G.A.3	Partition circles and rectangles into two, three, or four equal shares, describe the shares using the words halves, thirds, half of, a third of, etc., and describe the whole as two halves, three thirds, four fourths. Recognize that equal shares of identical wholes need not have the same shape.	12-4, 12-5

Correlations

Grade 3

◆ Major ▲ Supporting ○ Additional

	Standard Text		RM Lesson(s)
Operations and Algebraic Thinking			
◆	3.OA.A.1	Interpret products of whole numbers, e.g., interpret 5 × 7 as the total number of objects in 5 groups of 7 objects each.	3-1, 3-2, 3-6
◆	3.OA.A.2	Interpret whole-number quotients of whole numbers, e.g., interpret 56 ÷ 8 as the number of objects in each share when 56 objects are partitioned equally into 8 shares, or as a number of shares when 56 objects are partitioned into equal shares of 8 objects each.	3-4, 3-5, 3-6
◆	3.OA.A.3	Use multiplication and division within 100 to solve word problems in situations involving equal groups, arrays, and measurement quantities, e.g., by using drawings and equations with a symbol for the unknown number to represent the problem.	4-6, 5-7, 11-5
◆	3.OA.A.4	Determine the unknown whole number in a multiplication or division equation relating three whole numbers.	3-7, 4-6, 5-7, 11-5
◆	3.OA.B.5	Apply properties of operations as strategies to multiply and divide.	3-3, 5-1, 10-3
◆	3.OA.B.6	Understand division as an unknown-factor problem.	9-1
◆	3.OA.C.7	Fluently multiply and divide within 100, using strategies such as the relationship between multiplication and division (e.g., knowing that 8 × 5 = 40, one knows 40 ÷ 5 = 8) or properties of operations. By the end of Grade 3, know from memory all products of two one-digit numbers.	4-1, 4-2, 4-3, 4-4, 4-5, 5-2, 5-3, 5-4, 5-5, 5-6, 9-2, 9-3, 9-4, 9-5, 9-6, 9-7, 9-8, 9-9
◆	3.OA.D.8	Solve two-step word problems using the four operations. Represent these problems using equations with a letter standing for the unknown quantity. Assess the reasonableness of answers using mental computation and estimation strategies including rounding.	2-12, 10-4, 10-5, 10-6
◆	3.OA.D.9	Identify arithmetic patterns (including patterns in the addition table or multiplication table), and explain them using properties of operations.	2-5, 4-4, 10-2
Number and Operations in Base Ten			
○	3.NBT.A.1	Use place value understanding to round whole numbers to the nearest 10 or 100.	2-1, 2-2, 2-3
○	3.NBT.A.2	Fluently add and subtract within 1000 using strategies and algorithms based on place value, properties of operations, and/or the relationship between addition and subtraction.	2-3, 2-4, 2-6, 2-7, 2-8, 2-9, 2-10, 2-11
○	3.NBT.A.3	Multiply one-digit whole numbers by multiples of 10 in the range 10–90 (e.g., 9 × 80, 5 × 60) using strategies based on place value and properties of operations.	10-1

Standard Text		RM Lesson(s)
Number and Operations—Fractions		
◆ 3.NF.A.1	Understand a fraction $\frac{1}{b}$, with denominators 2, 3, 4, 6, and 8, as the quantity formed by 1 part when a whole is partitioned into b equal parts; understand a fraction $\frac{a}{b}$ as the quantity formed by a parts of size $\frac{1}{b}$.	7-2
◆ 3.NF.A.2	Understand a fraction with denominators 2, 3, 4, 6, and 8 as a number on a number line diagram. a. Represent a fraction $\frac{1}{b}$ on a number line diagram by defining the interval from 0 to 1 as the whole and partitioning it into b equal parts. Recognize that each part has size $\frac{1}{b}$ and that the endpoint of the part based at 0 locates the number $\frac{1}{b}$ on the number line. b. Represent a fraction $\frac{a}{b}$ on a number line diagram by marking off a lengths $\frac{1}{b}$ from 0. Recognize that the resulting interval has size $\frac{a}{b}$ and that its endpoint locates the number $\frac{a}{b}$ on the number line.	7-3, 7-6
◆ 3.NF.A.3	Explain equivalence of fractions with denominators 2, 3, 4, 6, and 8 in special cases, and compare fractions by reasoning about their size. a. Understand two fractions as equivalent (equal) if they are the same size, or the same point on a number line. b. Recognize and generate simple equivalent fractions, e.g., $\frac{1}{2} = \frac{2}{4}, \frac{4}{6} = \frac{2}{3}$). Explain why the fractions are equivalent, e.g., by using a visual fraction model. c. Express whole numbers as fractions, and recognize fractions that are equivalent to whole numbers. d. Compare two fractions with the same numerator or the same denominator by reasoning about their size. Recognize that comparisons are valid only when the two fractions refer to the same whole. Record the results of comparisons with the symbols >, =, or <, and justify the conclusions, e.g., by using a visual fraction model.	7-4, 7-5, 8-1, 8-2, 8-3, 8-4, 8-5, 8-6, 8-7
Measurement and Data		
◆ 3.MD.A.1	Tell and write time to the nearest minute and measure time intervals in minutes. Solve word problems involving addition and subtraction of time intervals in minutes, e.g., by representing the problem on a number line diagram.	12-5, 12-6
◆ 3.MD.A.2	Measure and estimate liquid volumes and masses of objects using standard units of grams (g), kilograms (kg), and liters (l). Add, subtract, multiply, or divide to solve one-step word problems involving masses or volumes that are given in the same units, e.g., by using drawings (such as a beaker with a measurement scale) to represent the problem.	12-1, 12-2, 12-3, 12-4

Standard Text			RM Lesson(s)
▲	3.MD.B.3	Draw a scaled picture graph and a scaled bar graph to represent a data set with several categories. Solve one- and two-step "how many more" and "how many less" problems using information presented in scaled bar graphs.	12-7, 12-8, 12-9
▲	3.MD.B.4	Generate measurement data by measuring lengths using rulers marked with halves and fourths of an inch. Show the data by making a line plot, where the horizontal scale is marked off in appropriate units—whole numbers, halves, or quarters.	12-10, 12-11
◆	3.MD.C.5	Recognize area as an attribute of plane figures and understand concepts of area measurement. a. A square with side length 1 unit, called "a unit square," is said to have "one square unit" of area, and can be used to measure area. b. A plane figure which can be covered without gaps or overlaps by *n* unit squares is said to have an area of *n* square units.	6-1
◆	3.MD.C.6	Measure areas by counting unit squares (square cm, square m, square in, square ft, and improvised units).	6-2
◆	3.MD.C.7	Relate area to the operations of multiplication and addition. a. Find the area of a rectangle with whole-number side lengths by tiling it, and show that the area is the same as would be found by multiplying the side lengths. b. Multiply side lengths to find areas of rectangles with whole-number side lengths in the context of solving real-world and mathematical problems and represent whole-number products as rectangular areas in mathematical reasoning. c. Use tiling to show in a concrete case that the area of a rectangle with whole-number side lengths a and $b + c$ is the sum of $a \times b$ and $a \times c$. Use area models to represent the distributive property in mathematical reasoning. d. Recognize area as additive. Find areas of rectilinear figures by decomposing them into non-overlapping rectangles and adding the areas of the non-overlapping parts, applying this technique to solve real world problems.	6-1, 6-3, 6-4, 6-5, 6-6
○	3.MD.D.8	Solve real-world and mathematical problems involving perimeters of polygons, including finding the perimeter given the side lengths, finding an unknown side length, and exhibiting rectangles with the same perimeter and different areas or with the same area and different perimeters.	11-1, 11-2, 11-3, 11-4

Standard Text		RM Lesson(s)
Geometry		
▲ 3.G.A.1	Understand that shapes in different categories (e.g., rhombuses, rectangles, and others) may share attributes (e.g., having four sides), and that the shared attributes can define a larger category (e.g., quadrilaterals). Recognize rhombuses, rectangles, and squares as examples of quadrilaterals, and draw examples of quadrilaterals that do not belong to any of these subcategories.	13-1, 13-2, 13-3, 13-4
▲ 3.G.A.2	Partition shapes into parts with equal areas. Express the area of each part as a unit fraction of the whole.	7-1, 7-2

Correlations

Grade 4

◆ Major ▲ Supporting ○ Additional

	Standard Text		RM Lesson(s)
Operations and Algebraic Thinking			
◆	4.OA.A.1	Interpret a multiplication equation as a comparison and represent verbal statements of multiplicative comparisons as multiplication equations, e.g., interpret $35 = 5 \times 7$ as a statement that 35 is 5 times as many as 7, and 7 times as many as 5.	4-1, 4-2
◆	4.OA.A.2	Multiply or divide to solve word problems involving multiplicative comparison, e.g., by using drawings and equations with a symbol for the unknown number to represent the problem, distinguishing multiplicative comparison from additive comparison (Example: 6 times as many vs. 6 more than).	4-2, 4-3, 4-4, 7-8
◆	4.OA.A.3	Solve multi-step word problems posed with whole numbers and having whole-number answers using the four operations, including problems in which remainders must be interpreted. Represent these problems using equations with a letter standing for the unknown quantity. Assess the reasonableness of answers using mental computation and estimation strategies including rounding.	3-1, 3-8, 3-9, 6-1, 6-2, 6-3, 6-4, 6-5, 6-6, 6-7, 6-8, 7-1, 7-2, 7-3, 7-4, 7-5, 7-6, 7-7, 13-7, 13-8, 13-9
▲	4.OA.B.4	Find all factor pairs for a whole number in the range 1–100. Recognize that a whole number is a multiple of each of its factors. Determine whether a given whole number in the range 1–100 is a multiple of a given one-digit number. Determine whether a given whole number in the range 1–100 is prime or composite.	5-1, 5-2, 5-3
○	4.OA.C.5	Generate a number or shape pattern that follows a given rule. Identify apparent features of the pattern that were not explicit in the rule itself.	5-4, 5-5, 5-6, 8-1, 8-2, 8-3
Number and Operations in Base Ten			
◆	4.NBT.A.1	Recognize that in a multi-digit whole number, a digit in one place represents ten times what it represents in the place to its right.	2-1, 8-4, 8-5
◆	4.NBT.A.2	Read and write multi-digit whole numbers using base-ten numerals, number names, and expanded form. Compare two multi-digit numbers based on meanings of the digits in each place, using >, =, and < symbols to record the results of comparisons.	2-1, 2-2, 2-3, 6-1, 6-2, 6-3, 6-4, 6-5, 6-6, 6-7
◆	4.NBT.A.3	Use place value understanding to round multi-digit whole numbers to any place.	2-4, 6-1, 6-2, 6-3, 6-4, 6-5, 6-6, 6-7, 7-1
◆	4.NBT.B.4	Fluently add and subtract multi-digit whole numbers, with sums less than or equal to 1,000,000, using the standard algorithm;	3-2, 3-3, 3-4, 3-5, 3-6, 3-7, 7-8

136 Content Guide

Standard Text		RM Lesson(s)
◆ 4.NBT.B.5	Multiply a whole number of up to four digits by a one-digit whole number, and multiply two two-digit numbers, using strategies based on place value and the properties of operations. Illustrate and explain the calculation by using equations, rectangular arrays, and/or area models.	6-1, 6-2, 6-3, 6-4, 6-5, 6-6, 6-7
◆ 4.NBT.B.6	Find whole-number quotients and remainders with up to four-digit dividends and one-digit divisors, using strategies based on place value, the properties of operations, and/or the relationship between multiplication and division. Illustrate and explain the calculation by using equations, rectangular arrays, and/or area models.	7-1, 7-2, 7-3, 7-4, 7-5, 7-6, 7-8
Number and Operations—Fractions		
◆ 4.NF.A.1	Explain why a fraction $\frac{a}{b}$ is equivalent to a fraction $\frac{(n \times a)}{(n \times b)}$ by using visual fraction models, with attention to how the number and size of the parts differ even though the two fractions themselves are the same size. Use this principle to recognize and generate equivalent fractions.	8-1, 8-2, 8-3, 8-4, 8-5
◆ 4.NF.A.2	Compare two fractions with different numerators and different denominators, e.g., by creating common denominators or numerators, or by comparing to a benchmark fraction such as $\frac{1}{2}$. Recognize that comparisons are valid only when the two fractions refer to the same whole. Record the results of comparisons with symbols >, =, or <, and justify the conclusions, e.g., by using a visual fraction model.	8-4, 8-5
◆ 4.NF.B.3	Understand a fraction $\frac{a}{b}$ with $a > 1$ as a sum of fractions $\frac{1}{b}$. a. Understand addition and subtraction of fractions as joining and separating parts referring to the same whole. b. Decompose a fraction into a sum of fractions with the same denominator in more than one way, recording each decomposition by an equation. Justify decompositions, e.g., by using a visual fraction model. c. Add and subtract mixed numbers with like denominators, e.g., by replacing each mixed number with an equivalent fraction, and/or by using properties of operations and the relationship between addition and subtraction. d. Solve word problems involving addition and subtraction of fractions referring to the same whole and having like denominators, e.g., by using visual fraction models and equations to represent the problem.	9-1, 9-2, 9-3, 9-4, 9-5, 9-6, 10-1, 10-2, 10-3, 10-4, 10-5, 10-6, 11-2, 11-5

Standard Text			RM Lesson(s)
◆	4.NF.B.4	Multiply a fraction by a whole number. (Denominators are limited to 2, 3, 4, 5, 6, 8, 10, 12, and 100.) a. Understand a fraction $\frac{a}{b}$ as a multiple of $\frac{1}{b}$. b. Understand a multiple of $\frac{a}{b}$ as a multiple of $\frac{1}{b}$, and use this understanding to multiply a fraction by a whole number. c. Solve word problems involving multiplication of a fraction by a whole number, e.g., by using visual fraction models and equations to represent the problem.	11-1, 11-2, 11-3, 11-4, 11-5
◆	4.NF.C.5	Express a fraction with denominator 10 as an equivalent fraction with denominator 100, and use this technique to add two fractions with respective denominators 10 and 100.	12-1, 12-4
◆	4.NF.C.6	Use decimal notation for fractions with denominators 10 or 100.	12-2
◆	4.NF.C.7	Compare two decimals to hundredths by reasoning about their size. Recognize that comparisons are valid only when the two decimals refer to the same whole. Record the results of comparisons with the symbols >, =, or <, and justify the conclusions, e.g., by using a visual model.	12-3
Measurement and Data			
▲	4.MD.A.1	Know relative sizes of measurement units within one system of units including: ft, in; km, m, cm; kg, g; lb, oz.; l, ml; hr, min, sec. Within a single system of measurement, express measurements in a larger unit in terms of a smaller unit. (Conversions are limited to one-step conversions.) Record measurement equivalents in a two-column table.	13-1, 13-2, 13-3, 13-4
▲	4.MD.A.2	Use the four operations to solve word problems involving distances, intervals of time, liquid volumes, masses of objects, and money, including problems involving whole numbers and/or simple fractions (addition and subtraction of fractions with like denominators and multiplying a fraction times a fraction or a whole number), and problems that require expressing measurements given in a larger unit in terms of a smaller unit. Represent measurement quantities using diagrams such as number line diagrams that feature a measurement scale.	12-5, 13-2, 13-5, 13-6
▲	4.MD.A.3	Apply the area and perimeter formulas for rectangles in real-world and mathematical problems.	13-7, 13-8, 13-9
▲	4.MD.B.4	Make a line plot to display a data set of measurements in fractions of a unit ($\frac{1}{2}, \frac{1}{4}, \frac{1}{8}$). Solve problems involving addition and subtraction of fractions by using information presented in line plots.	13-10, 13-11

Standard Text		RM Lesson(s)
○ 4.MD.C.5	Recognize angles as geometric shapes that are formed wherever two rays share a common endpoint and understand concepts of angle measurement.	14-2, 14-3
	a. An angle is measured with reference to a circle with its center at the common endpoint of the rays, by considering the fraction of the circular arc between the points where the two rays intersect the circle.	
	b. An angle that turns through $\frac{1}{360}$ of a circle is called a "one-degree angle," and can be used to measure angles.	
	c. An angle that turns through n one-degree angles is said to have an angle measure of n degrees.	
○ 4.MD.C.6	Measure angles in whole-number degrees using a protractor. Sketch angles of specified measure.	14-3
○ 4.MD.C.7	Recognize angle measure as additive. When an angle is decomposed into non-overlapping parts, the angle measure of the whole is the sum of the angle measures of the parts. Solve addition and subtraction problems to find unknown angles on a diagram in real-world and mathematical problems, e.g., by using an equation with a letter for the unknown angle measure.	14-5, 14-6
Geometry		
○ 4.G.A.1	Draw points, lines, line segments, rays, angles (right, acute, obtuse), and perpendicular and parallel lines. Identify these in two-dimensional figures.	14-1, 14-2, 14-4, 14-7, 14-8
○ 4.G.A.2	Classify two-dimensional figures based on the presence or absence of parallel or perpendicular lines, or the presence or absence of angles of a specified size. Recognize right triangles as a category, and identify right triangles.	14-7, 14-8
○ 4.G.A.3	Recognize a line of symmetry for a two-dimensional figure as a line across the figure such that the figure can be folded along the line into matching parts. Identify line-symmetric figures and draw lines of symmetry.	14-9, 14-10

Correlations

Grade 5

◆ Major   ▲ Supporting   ○ Additional

	Standard Text		RM Lesson(s)
Operations and Algebraic Thinking			
○	5.OA.A.1	Use parentheses or brackets in numerical expressions and evaluate expressions with these symbols.	14-1, 14-2, 14-3
○	5.OA.A.2	Write simple expressions that record calculations with whole numbers, fractions and decimals, and interpret numerical expressions without evaluating them.	14-1, 14-2
○	5.OA.B.3	Generate two numerical patterns using two given rules. Identify apparent relationships between corresponding terms. Form ordered pairs consisting of corresponding terms from the two patterns and graph the ordered pairs on a coordinate plane.	14-4, 14-5, 14-6
Number and Operations in Base Ten			
◆	5.NBT.A.1	Recognize that in a multi-digit number, a digit in one place represents 10 times as much as it represents in the place to its right and $\frac{1}{10}$ of what it represents in the place to its left.	3-1, 3-2
◆	5.NBT.A.2	Explain patterns in the number of zeros of the product when multiplying a number by powers of 10 and explain patterns in the placement of the decimal point when a decimal is multiplied or divided by a power of 10. Use whole-number exponents to denote powers of 10.	5-1, 5-2, 6-1, 8-1
◆	5.NBT.A.3	Read, write, and compare decimals to thousandths. a. Read and write decimals to thousandths using base-ten numerals, number names, and expanded form. b. Compare two decimals to thousandths based on meanings of the digits in each place, using >, =, and < symbols to record the results of comparisons.	3-3, 3-4
◆	5.NBT.A.4	Use place value understanding to round decimals to any place.	3-5
◆	5.NBT.B.5	Fluently multiply multi-digit whole numbers using the standard algorithm.	5-3, 5-4, 5-5, 5-6, 5-7
◆	5.NBT.B.6	Find whole-number quotients of whole numbers with up to four-digit dividends and two-digit divisors, using strategies based on place value, the properties of operations, subtracting multiples of the divisor, and/or the relationship between multiplication and division. Illustrate and/or explain the calculation by using equations, rectangular arrays, area models, or other strategies based on place value.	7-1, 7-2, 7-3, 7-4, 7-5, 7-6, 7-7
◆	5.NBT.B.7	Add, subtract, multiply, and divide decimals to hundredths, using concrete models or drawings and strategies based on place value, properties of operations, and/or the relationship between addition and subtraction; justify the reasoning used with a written explanation.	4-1, 4-2, 4-3, 4-4, 4-5, 4-6, 4-7, 4-8, 6-2, 6-3, 6-4, 6-5, 6-6, 8-2, 8-3, 8-4, 8-5, 8-6

140   Content Guide

Standard Text		RM Lesson(s)
Number and Operations—Fractions		
◆ 5.NF.A.1	Add and subtract fractions with unlike denominators (including mixed numbers) by replacing given fractions with equivalent fractions in such a way as to produce an equivalent sum or difference of fractions with like denominators.	9-2, 9-3, 9-4, 9-5, 9-6, 9-7, 9-8, 9-9
◆ 5.NF.A.2	Solve word problems involving addition and subtraction of fractions referring to the same whole, including cases of unlike denominators, e.g., by using visual fraction models or equations to represent the problem. Use benchmark fractions and number sense of fractions to estimate mentally and assess the reasonableness of answers.	9-1
◆ 5.NF.B.3	Interpret a fraction as division of the numerator by the denominator ($\frac{a}{b} = a \div b$). Solve word problems involving division of whole numbers leading to answers in the form of fractions or mixed numbers, e.g., by using visual fraction models or equations to represent the problem.	11-1
◆ 5.NF.B.4	Apply and extend previous understandings of multiplication to multiply a fraction or whole number by a fraction. a. Interpret the product ($\frac{a}{b}$) × q as a parts of a partition of q into b equal parts; equivalently, as the result of a sequence of operations, a × q ÷ b. b. Find the area of a rectangle with fractional side lengths by tiling it with unit squares of the appropriate unit fraction side lengths and show that the area is the same as would be found by multiplying the side lengths. Multiply fractional side lengths to find areas of rectangles and represent fraction products as rectangular areas.	10-1, 10-2, 10-3, 10-4, 10-5, 10-6, 10-7
◆ 5.NF.B.5	Interpret multiplication as scaling (resizing), by: a. Comparing the size of a product to the size of one factor on the basis of the size of the other factor, without performing the indicated multiplication. b. Explaining why multiplying a given number by a fraction greater than 1 results in a product greater than the given number (recognizing multiplication by whole numbers greater than 1 as a familiar case); explaining why multiplying a given number by a fraction less than 1 results in a product smaller than the given number; and relating the principle of fraction equivalence $\frac{a}{b} = \frac{(n \times a)}{(n \times b)}$ to the effect of multiplying $\frac{a}{b}$ by 1.	10-8
◆ 5.NF.B.6	Solve real-world problems involving multiplication of fractions and mixed numbers, e.g., by using visual fraction models or equations to represent the problem.	10-9

Standard Text		RM Lesson(s)
◆ 5.NF.B.7	Apply and extend previous understandings of division to divide unit fractions by whole numbers and whole numbers by unit fractions.	11-3, 11-4, 11-5, 11-6, 11-7
	a. Interpret division of a unit fraction by a non-zero whole number and compute such quotients.	
	b. Interpret division of a whole number by a unit fraction and compute such quotients.	
	c. Solve real-world problems involving division of unit fractions by non-zero whole numbers and division of whole numbers by unit fractions, e.g., by using visual fraction models and equations to represent the problem.	
Measurement and Data		
▲ 5.MD.A.1	Convert among different-sized standard measurement units within a given measurement and use these conversions in solving multi-step, real-world problems (e.g., convert 5 cm to 0.05 m; 9 ft to 108 in).	12-1, 12-2, 12-3
▲ 5.MD.B.2	Make a line plot to display a data set of measurements in fractions of a unit ($\frac{1}{2}, \frac{1}{4}, \frac{1}{8}$). Use operations on fractions for this grade to solve problems involving information presented in line plots.	12-4, 12-5
◆ 5.MD.C.3	Recognize volume as an attribute of solid figures and understand concepts of volume measurement.	2-1, 2-2
	a. A cube with side length 1 unit, called a "unit cube," is said to have "one cubic unit" of volume, and can be used to measure volume.	
	b. A solid figure which can be packed without gaps or overlaps using n unit cubes is said to have a volume of n cubic units.	
◆ 5.MD.C.4	Measure volumes by counting unit cubes, using cubic cm, cubic in, cubic ft, and improvised units.	2-2

Standard Text		RM Lesson(s)
◆ **5.MD.C.5**	Relate volume to the operations of multiplication and addition and solve real-world and mathematical problems involving volume.	2-3, 2-4, 2-5
	a. Find the volume of a right rectangular prism with whole-number side lengths by packing it with unit cubes, and show that the volume is the same as would be found by multiplying the edge lengths, equivalently by multiplying the height by the area of the base. Represent threefold whole-number products as volumes, e.g., to represent the associative property of multiplication.	
	b. Apply the formulas $V = l \times w \times h$ and $V = b \times h$ for rectangular prisms to find volumes of right rectangular prisms with whole-number edge lengths in the context of solving real-world and mathematical problems.	
	c. Recognize volume as additive. Find volumes of solid figures composed of two non-overlapping right rectangular prisms by adding the volumes of the non-overlapping parts, applying this technique to solve real-world problems.	
Geometry		
○ **5.G.A.1**	Use a pair of perpendicular number lines, called axes, to define a coordinate system, with the intersection of the lines (the origin) arranged to coincide with the 0 on each line and a given point in the plane located by using an ordered pair of numbers, called its coordinates. Understand that the first number in the ordered pair indicates how far to travel from the origin in the direction of one axis, and the second number in the ordered pair indicates how far to travel in the direction of the second axis, with the convention that the names of the two axes and the coordinates correspond (e.g., *x*-axis and *x*-coordinate, *y*-axis and *y*-coordinate).	13-1, 13-2
○ **5.G.A.2**	Represent real-world and mathematical problems by graphing points in the first quadrant of the coordinate plane and interpret coordinate values of points in the context of the situation.	13-3
○ **5.G.B.3**	Understand that attributes belonging to a category of two-dimensional figures also belong to all subcategories of that category.	13-4, 13-6
○ **5.G.B.4**	Classify two-dimensional figures in a hierarchy based on properties.	13-4, 13-5, 13-6

Correlations **143**